STEPHEN HAWKING
ON TRIAL

CONFRONTING THE BIG BANG

PIERRE ST. CLAIR

Open Mind
Publishers

Book layout by Natasa Marovic
Cover design by Peter Cover

Published simultaneously in the United States of America
and Canada by Open Mind Publishers

Library of Congress Cataloging-in-Publication data
available from the Publisher

Reviews

A well thought out argument against the dangers of arrogant science. Mr. St. Clair takes one of the biggest names in modern science to task and puts the inconsistencies on trial. This book is an interesting read and reminds the reader that science should never be thought of as "settled."
S. J. Thompson – Northern California

This is the first book that has assembled the origins of cosmology as evidence like opposing counsels. Readers can weigh testimonies as a judge who demands rigorous arguments and proofs. St. Clair has the detached presence of mind to act as both scales of justice.
D. M. Shapiro – Los Angeles filmmaker

The author in no uncertain terms establishes the need to highlight the word theoretical in theoretical physics. Stephen Hawking is given due respect for his achievements but some serious challenges are raised which indicate the need for Hawking to reduce the confidence and certainty with which he propagates his theories. It's clear the author has put considerable research into the work.
A. Gallagher – New Zealand

Stephen Hawking On Trial is compelling reading for anyone with an interest in physics and cosmology. It may challenge your beliefs, or maybe not, but it's a fascinating read in either case.
J. Chambers – Amazon Top 10 Reviewer

"Hawking and a few others are doing a disservice to science by exaggerating what we know and therein confusing the public. This book sets the record straight."
D. Friedmann – Canadian author

This is the best science book that I've read in quite a while; a science book that views Hawking's work and statements critically in a clear concise manner. I've been aware of the criticisms of Hawking's work...
H. Lipman– American author

Dedication

To my teacher who opened my darkened eyes
with the torchlight of knowledge

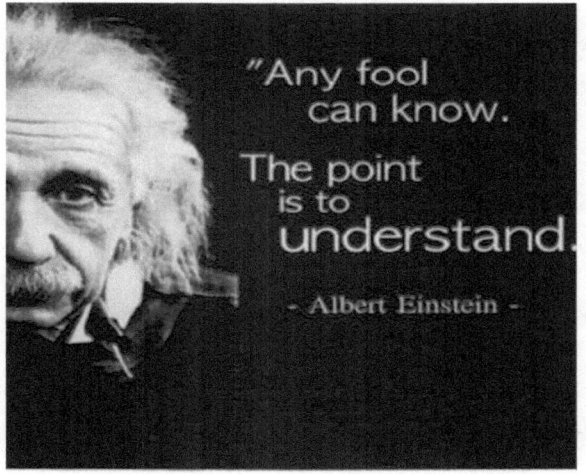

Table of Contents

INTRODUCTION

There are various anomalies within our cosmology about which many astrophysicists disagree. The renowned American astronomer, Dr. Carl Sagan, in a more conciliatory mood remarked that, "There are many hypotheses in science which are wrong. That's perfectly all right; they're the aperture to finding out what's right."

In other words, science is an ongoing exploration to discover the mysteries of the universe. In this book I examine the latest theory/explanation of Dr. Stephen Hawking on the origin of our cosmos.

Unfortunately, many cosmological theories are never devised as an experiment under controlled laboratory conditions. Rather, modern cosmologists sit in an office and devise their theories on paper, or computer, using mathematics.

Truth be told, both mathematicians and physicists candidly wonder whether their mathematical equations factually represent the real world.

Meanwhile, media reports assure the public that science is close to unlocking the secrets of the universe leaving only minor details to be explained. Similar, in some respects, to how the media forecast the victory of Hillary Clinton.

I do admit that I also blindly accepted reports about scientific advances, especially regarding cosmology and the origin of the universe. However, my in-depth study of the nature and evolution of the cosmos revealed contradictory explanations, although each was presented as reliable and scientific.

My job as an investigative journalist directed that I dig deeper to ascertain which reports were accurate. My probing questions uncovered "hidden" details behind the headlines.

Let's take a look at the bedrock concepts of genuine science under the microscope. First, is the testability of the hypothesis. Second, is the verification of data that supports the hypothesis. Everything must be verified either by observation or experimentation.

The true methodology of science requires carefully analyzing ideas encompassed in a premise. Every hypothesis must be tested rigorously, including attempts to disprove it. Falsifiability is considered a strength.[1]

Research, challenges, and critiques are integral factors of genuine science. When hypotheses tested via known scientific principles are not proven false, they can then be elevated to the status of a theory.

We should be clear that if observational data cannot be demonstrated there should be no claim of a theory. Even withpredictions from mathematics, what remains is simply an uncorroborated hypothesis. This is the basis of Edwin Hubble's book, *The Observational Approach to Cosmology.*

Again, modern cosmology lacks experiments under controlled laboratory conditions. Thus, 21st century astrophysicists devise their theories in their office twiddling numbers on a laptop.

Physics in 1915

A hundred years ago, the modern creation story was altogether different. In all the best universities our Milky Way galaxy was taught to be the entire universe.

Looking back at this understanding from our present perspective it appears quaint, even humorous. Moreover, the cosmos was believed to be eternal and static. This was the teaching for over three centuries, accepted by all renowned scientists including Albert Einstein.

This made me question, how would 22nd century scientists view today's explanation?

As we discover more of nature's secrets, the universe has become confusingly complex. Honest scientists willingly admit that the more data our instruments supply the more difficult it is to make sense of it.

In this connection, astronomer Sir Fred Hoyle commented:

> "The whole history of science shows that each generation finds the universe to be stranger than the preceding generation ever conceived it to be."[2]

A major discovery was the fact that the universe operated according to strict natural laws that could be understood via mathematics.

Galileo put it this way:

"The great book of nature can be read only by those who know the language in which it was written. And this language is mathematics."[3]

The British astronomer Sir James Jeans stated it more elegantly:

"The universe appears to have been designed by a pure mathematician."[4]

Of course, today the term "design" has pejorative connotations in the cosmological community. Nonetheless, have we discovered an invisible sub-text of nature? Is there a cosmic code by which the universe obeys mathematical logic? If mathematics is the language of the universe, how did that language come into existence? Or, have we invented it?

Historically, Sir Isaac Newton did invent the mathematics he needed to describe the laws he discovered. He called his math the theory of fluxions. Today it's known as calculus. He accepted it as an underlying fundamental principle by which the universe was governed – the key to a cosmic code.

Keep in mind that Newton did not have an explanation for the causes of gravity, inertia, and centrifugal force, for which he became famous. But he did have a talent for figuring out the mathematical relationships of these natural forces. Since then, physicists have used mathematics to interpret the Laws of Nature and predict the outcome of natural phenomena.

THE MATH OF PHYSICS

Today, theoretical physicists create mathematical formulations to model real world situations. They fiddle with figures in mathematical equations to decipher what would happen in real world conditions without performing the experiment.

Does this method work in all cases? What if the equation is not an accurate representation?

Mario Livio, head of the science division of the Hubble Space Telescope, explains how mathematics simply fills a role that cosmologists need to describe events and concepts.

> "The success of pure mathematics turned into applied mathematics, in this picture, merely reflects an overproduction of concepts, from which physics has selected the most adequate for its needs – a true survival of the fittest. After all, 'inventionists' would point out, Godfrey H. Hardy was always proud of having 'never done anything useful.'
>
> "This opinion of mathematics is apparently espoused also by Marilyn vos Savant, the World Record holder in IQ – an incredible 228. She is quoted as having said, 'I'm beginning to think simply that mathematics can be invented to describe anything, and matter is no exception.'"[5]

Savant, an American author, lecturer, and playwright, is listed in the *Guinness Book of World Records* under "Highest IQ."

Godfrey Hardy was a prominent English mathematician known for achievements in number theory and mathematical analysis. So these opinions come from renowned experts.

We may question why nature is understood by mathematics. Why do the equations actually work? The fact that nature seems to conform efficiently to mathematical formulations remains unexplained by Stephen Hawking.

Joseph Needham points out that mathematics had already encroached upon physics by the turn of the 20th century. It was invented to satisfy the requirements of the day.

> "The mathematisation of physics...is continually growing and physics is becoming more and more dependent upon the fate of mathematics....This special mathematics has for the greater part been created by the physicists themselves, for ordinary mathematics is unable to satisfy the requirements of present day physics."[6]

Physicist Herbert Dingle is quite clear that mathematics can substantiate whatever is required.

> "...in the language of mathematics we can tell lies as well as truths, and within the scope of mathematics itself there is no possible way of telling one from the other. We can distinguish them only by experience or by reasoning outside the mathematics, applied to the possible relation between the mathematical solution and its supposed physical correlate."[7]

I'm shocked that mathematics can be manipulated according to the subjective intention of a theorist, thus rendering it unreliable and in some cases completely suspect.

Science philosopher, Karl Popper, establishes that physics by mathematics does not correlate to real world situations.

> "Properly understood, a mathematical hypothesis does not claim that anything exists in nature which corresponds to it...It erects, as it were, a fictitious mathematical world behind that of appearance, but without the claim that this world exists. [It is] to be

regarded only as a mathematical hypothesis, and not as anything really existing in nature."[8]

Does the mathematics of Einstein represent what really exists in nature? Or does the math merely save appearances? For example, in Relativity Theory is it just a case of saying $4 + 4 = 8$ when in reality the correct equation is $5 + 3 = 8$?

Morris Kline is a professor of mathematics at New York University and the Courant Institute. He is highly critical of mathematics and its applications to science. Here is the shocking analysis from an insider.

> "It is now apparent that the concept of a universally accepted, infallible body of reasoning – the majestic mathematics of 1800 and the pride of man – is a grand illusion. Uncertainty and doubt concerning the future of mathematics have replaced the certainties and complacency of the past.

> "The disagreements about the foundations of the "most certain" science are both surprising and, to put it mildly, disconcerting. The present state of mathematics is a mockery of the hitherto deep-rooted and widely reputed truth and logical perfection of mathematics. The disagreements concerning what correct mathematics is and the variety of differing foundations affect seriously not only mathematics proper but most vitally physical science."[9]

Clearly, our lucid understanding of mathematics from the past has been usurped by the probability and uncertainty of modern physics. Thomas Van Flandern points out that, "Mathematics should be used to describe the operation of models, not to build them."

"...equations cannot be made to substitute for the concepts which underlie them. And equations are generally blind to limitations of range and physical constraints. They are too general, and simply lack the sort of specificity that true, intuitive understanding demands. Every equation has a domain of applicability – usually the range of the observations and little, if anything, more...

"If an equation can be extrapolated outside its domain and gives a singularity (basically, a zero divisor), that singularity does not exist in nature; instead, the model needs modification. Up to now this rule has always proved true. But advocates of "black holes" in the universe would have us believe that the equations which predict them can be relied upon far outside the domain of the observations used to derive those equations."[10]

The mathematical singularity does not exist in nature? Physicist Michael Duff is quite candid about this:

"Well, the question we often ask ourselves as we work through our equations: 'Is this just fancy mathematics, or is it describing the real world?'"[11]

In his comments on the issue, the celebrated British philosopher Bertrand Russell is particularly sarcastic.

"Pure mathematics consists entirely of assertions to the effect that if such and such a proposition is true of anything then such and such another proposition is true of that thing. It is essential not to discuss whether the first proposition is really true, and not to mention what the anything is, of which it is supposed to be true.

Both of these points would belong to applied mathematics... Thus mathematics may be defined as the subject in which we never know what we are talking about, nor what we are saying is true."[12]

Dr. Lee Smolin represents the Perimeter Institute for Theoretical Physics. He claims that the mathematisation of physics has resulted in the reduction of the cosmos to a mathematical entity. This has not only confused physicists but accounts for their worst and most distracting assertions.

Other physicists don't agree. Max Tegmark of M.I.T has a published book called *Our Mathematical Universe* wherein he states his belief that our physical reality is a mathematical structure. I say "belief" because there is no data to substantiate this idea. His colleagues affectionately call him Mad Max.

Of course, foremost among celebrity astrophysicists is Professor Stephen Hawking of Cambridge University. Unexpectedly, I found a lot of anomalous material in his latest explanation of the origin of the universe.

I began to wonder if his latest theory would stand up to the rigorous procedures in a court of law where facts and hard data are accepted as evidence, never conjecture. The result of my research is this book wherein a mock trial calls Stephen Hawking to the witness stand.

The trial begins with a replay of a "Larry King Live" television interview. Professor Hawking has to support his latest origin of the universe hypothesis.

Next, our mock prosecuting attorney presents a counterbalance to Hawking's ideas by reviewing a scientific panel discussing his new cosmology book *The Grand Design*.

Succeeding chapters present a detailed analysis of Hawking's "grand design" creation hypothesis, which he calls a theory. Various scientists challenge his points as witnesses for the prosecution. The prosecutor identifies and evaluates facts and faults in Hawking's account.

You, dear reader, will sit on the jury and hear the arguments both pro and con. After hearing the evidence you will decide for yourself the legitimacy of Stephen Hawking's scientific explanations.

We all have our own opinions about the "creation conundrum" and my job is simply to inform, not transform, what people have the right to know.

In this regard I may point out that some people retain a vested interest in their opinion of what is true, in spite of evidence to the contrary. That's clearly confirmed in book reviews.

It's humorous to see people post insulting reviews having never purchased the book. They are unaware that celebrated physicists are expressing their conclusions. Others, judge it to be religious based because the topic is broached in a TV interview.

Some people post 1 star reviews filled with ridicule after a cursory glance without having read the entire book.Such "reviewers" tend to vilify writers with a differing outlook regardless of the preponderance of evidence. Some even resort to hate mail. Of course, these people would never be chosen to sit on a jury.

Impartial and forthright readers, however, will appreciate that this book is a treatise on modern cosmology and nothing else.

1
On the Witness Stand

There are many medical practitioners. Among them, some are just quacks. That doesn't mean the entire field of medicine is a hoax. I raise this point because when we come across a quack doctor practicing medicine, we want to expose him. It's a public service to protect people from harm and exploitation by quack treatment. This is logical and responsible.

The situation is analogous in science. If a physicist manufactures a theory which purports to explain everything, then it's natural to examine his claim to ascertain its validity or its quackery. If quack it must be exposed and broadcast to the public. Thus, the uninformed public becomes the informed public. In this way they are protected from being duped.

Our mock trial begins with the prosecutor and the defense attorney presenting their opening statements. The defense introduces Stephen Hawking as a Professor of Mathematics at the University of Cambridge for 30 years; a theoretical physicist; the 2009 recipient of the Presidential Medal of Freedom; the author of many bestselling books, as well as the co-author of *The Grand Design* with Leonard Mlodinow.

"There is no doubt that professor Hawking is a man with outstanding credentials," he asserts, "and thus his opinions must be accepted as scientific and informed."

The prosecutor rises and does not disagree. "Hawking's career is impressive, indeed. Doubtless, he's a fine man who has overcome tremendous adversity. I saw the movie. But, his reputation is not on trial – only his hypotheses. Let the facts speak for themselves and allow the jury to come to its own conclusion."

He begins by setting up a screen for a playback of a "Larry King Live" interview originally televised on September 10, 2010.

"Professor Hawking's new book," he informs the jury, "purports to explain how and why the universe was created. It's the latest explanation of the modern creation story in a long line of various accounts." The video begins with this first question.

Larry King: "You say that science can explain the universe without the need for a creator. What is that explanation? Why is there something instead of nothing?"

Stephen Hawking: "Gravity and quantum theory cause universes to be created spontaneously out of nothing."

After each answer, the prosecutor pauses the video to comment.

"He states that gravity and quantum theory create universes. That's his reply, concise and to the point. Logically, if gravity and quantum theory are the causing agency then they must have existed prior to the creation of our cosmos. Where did *they* come from?

"Moreover, gravity and quantum theory are surely something, so how did the universe get created from nothing? No clue is given. The contradiction is left hanging.

"This response seems to mimic the traditional story that God created the universe spontaneously out of nothing. Gravity and quantum theory have now replaced the supernatural creator of traditional creation stories. Hawking has borrowed the traditional answer – that there was a cause, even in the 'nothingness' – but gives the credit to gravity and quantum theory.

"Ladies and gentlemen of the jury, gravity is an attractive force between objects. Did gravity exist before the universe, before there were forces and objects in existence? The standard answer given by physicists is via mathematics. They simply equate matter equal to zero in the equations defining gravitational force before the Big Bang.

"Please think about this for a moment. By equating matter equal to zero, have I explained anything about gravity? Notwithstanding this trivial solution, the question still remains how gravity *already* existed before existence came into being? Furthermore, if gravity can create objects out of nothing why do we not see that happening today, in our time?

"Can Einstein's equations really define and control the universe? As we shall learn later, Stephen Hawking declares in his *Grand Design*book that General Relativity equations break down at the origin of the cosmos. Therefore, GR theory can't be used to predict anything prior to the beginning of the universe. It can only be useful in explanations how it evolved afterwards.

"Besides this, if gravity and quantum theory existed before time and before space, then where were they located? Were they in another dimension beyond our known space, beyond the realm of nothing? What's more, Professor Hawking has avoided the question, *Why is there something rather than nothing?*"

With a quick tap the prosecutor starts the video again. Larry King doesn't appear to be satisfied with Hawking's answer. He wants to know how gravity and quantum theory came to be. So he tries again from another perspective.

LK: "You write that because there is a law such as gravity, the universe can and will create itself from nothing. Can you tell me how that law came into existence?"

SH: "Gravity is a consequence of M-theory which is the only possible unified theory. It is like saying, why is $2 + 2 = 4$?"

Turning to the jury our prosecutor exclaims, "There you have it. You heard the answer directly from Stephen Hawking. Now you know how the law of gravity came into existence. Or do you? I didn't get an answer I could write on a physics exam.

"Factually, $2 + 2 = 4$ is due to the values we assign to 2 and 4. Apparently, Hawking assigns values to gravity and M-theory to justify his conclusion, but what are those values? There was no conclusive answer. He simply added a step by saying; *Gravity is a consequence of M-theory*. How many of you jurors know anything about M-theory?

"We can't buy that the law of gravity is a consequence of M-theory if we don't know what that theory is. If it explains the consequence of the law of gravity and the universe, people want to hear that explanation. That's why Larry King is asking.

"Will the uninformed public think that M-theory is Hawking's theory? A simple Wikipedia search reveals more than the respected professor told us. It's a recent idea proposed by mathematician Edward Witten, who states that his M-theory is only an untested hypothesis:

> "M-theory (and string theory) has been criticized for lacking predictive power or being untestable. Further work continues to find mathematical constructs that join various surrounding theories. However, the tangible success of M-theory can be questioned, given its current incompleteness and limited predictive power." [Wikipedia]

"Clearly, Witten downplays his own hypothesis saying it has *limited predictive power*. Yet somehow Hawking can predict that the whole universe will arise from it?

"Moreover, did you notice that Hawking minimized the merit of King's question? *It is like saying why is 2 + 2 = 4?* In other words, the universe was created spontaneously out of nothing so it simply is what it is – just like 2 + 2 = 4.

"But, wait a minute. Everything else is what it is, too! What makes this a satisfactory answer? Now this logic will allow any religious extremist to give the same self-righteous reply: 'The universe is a consequence of God. He holds the key to the only possible unified theory. It's like saying why is 2 + 2 = 4? God is what he is. Case closed. End of story.'

"Any quack can now use the same logic as Hawking. It appears we are expected to accept on faith, while the public is left in the dark. If such an excuse for logic is considered valid for science,

16

it should also be valid for religion. What's good for the goose is good for the gander.

"Hawking avoided an answer because he implied it's meaningless to ask this question. It's like asking, why is 2 + 2 = 4? But when asked for the origin of gravity can you write, 'gravity is a consequence of M-theory' on your physics exam? This doesn't satisfy me, what to speak of my physics professor.

"Perhaps in his own mind Hawking thinks he's given a definitive reply. Yet, other physicists do not subscribe to such a consequence: *M-theory has been criticized for lacking predictive power or being untestable.*We have yet to hear anything substantial to validate that Stephen Hawking has solved the mystery of creation."

King's next question goes right to the point many in the audience most likely want to hear.

LK: "Do you believe in God?"

SH: "God may exist, but science can explain the universe without the need for a creator."

With this response, Hawking reveals a lot more about his belief system than he probably intended.

"His personal belief," says our prosecutor, "is that God didn't create the universe because quantum theory and the law of gravity did. But we need evidence in science, right? No hint of data is presented in this interview. He expects us to accept on faith, just as theologians do."

King patiently continues his questions.

LK: "Your book has stirred a lot of controversy. Why do you think people react so strongly to your contention that it is not necessary to invoke God to explain the creation of the universe?"

SH: "Science is increasingly answering questions that used to be the province of religion."

LK: "One of your colleagues out of Cambridge said that science provides us with the narrative as to how existence may happen, but theology addresses the meaning of the narrative. How do you respond to that?"

SH: "The scientific account is complete. Theology is unnecessary."

Again the video is paused. The prosecutor holds up a large graph for the jury to examine.

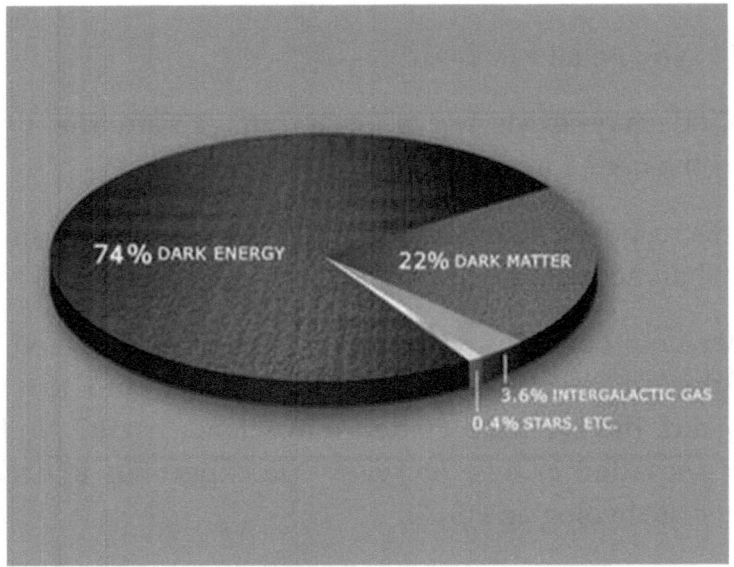

Credit: NASA

"Clearly, Hawking does not agree with his Cambridge colleague. But in this exchange he also contradicts NASA physicists by claiming, *the scientific account is complete.* Complete means there is nothing more to discover. Game over. End of story.

"We can see that scientists acknowledge a full 96 percent of the stuff of the universe is unknown. When cosmologists use the word "dark" it means something can't be seen or they have no idea what it is. Of course, they may harbor high hopes they can identify it soon.

"Yet for Stephen Hawking the scientific account is complete when only 4 percent of the matter and energy of the universe is known to science? A strange sense of completeness, indeed.

"And when the remaining 96 percent is finally understood, will Dr Hawking admit he was premature in claiming completeness today? This contradiction could well be a source of amusement 50 years from now; a joke among cosmologists in 2065. And as science advances into the 22nd century Hawking's Grand Design theory might become as comical as Henry Ford's belief that his horseless carriage was the pinnacle of human engineering."

Back to the video replay. Seeing that he's getting nowhere, King changes the subject to give the interview a human interest angle. Before closing the interview, however, he returns to the topic of the book.

LK: "What do you most hope people take away from your new book *The Grand Design?* What is the most important point in the book?"

SH: "That science can explain the universe, and that we don't need God to explain why there is something rather than nothing, or why the laws of nature are what they are."

Again the video is paused.

"By asserting that, *Science can explain the universe,* Hawking needs to give that explanation. But he hasn't given a definitive explanation. He presented no data, no evidence, no mention of dark matter or dark energy, or that 96 percent of the universe is unknown. Hawking sounded like a fundamentalist scientist lacking rational explanations.

"By saying, *we don't need God to explain why there is something rather than nothing,* he makes it sound less like a scientific explanation and more like a religious explanation. We still don't have a scientific answer I can wager my house on.

"Jurors, you may question whether Hawking is simply bluffing Larry King. Does it appear like he's borrowing religious doctrine and mocking theology? Remember, he disagreed with his theological colleague at Cambridge.

"Is he starting to sound like one of those know-it-all guys you meet in college that nobody likes?"

At this point the defense attorney jumps up. "Objection, your honor, this is a presumption on behalf of my colleague meant to sway the jury in a particular way."

The judge agrees. "Sustained! Facts are facts and it's irrelevant how or by whom we receive them. Jurors, disregard that last statement." He bangs his gavel to emphasize his authority.

"Nevertheless," the prosecutor contends, "without data to back up his assertion, Hawking does come across like a religious extremist beating the drum for his own religion – Scientism.

"Even the name of his book *The Grand Design* smacks of the designer that the Intelligent Design movement tries to establish. But now it appears that Hawking wants to be the intelligent designer. He invokes M-theory over and over with no mention of Edward Witten who rightly deserves credit as the originator of M-theory."

With a light tap of a finger the video playback continues.

SH: "According to M-theory ours is not the only universe. Instead, M-Theory predicts that a great many universes were created out of nothing. Their creation does not require the intervention of some supernatural being or God. Rather, these multiple universes arise naturally from physical law. They are a prediction of science."

The prosecution persists. "Well, once again Hawking claims predictive power from M-theory, when the originator of the theory states it lacks predictive power. We're hearing the same slogan with zero explanation from the scientific side. Is he reading from a script, or towing the party line?

"How did the physical laws come into being first, so that multiple universes could arise from them naturally? Professor Hawking has again added a sneaky step from nothing to something.

"It's like that old conjurer's trick. Pull a rabbit out of a hat, or a coin from a child's ear, to give the impression it mysteriously appeared from nowhere. It's the illusionist's stock of trade, betting that the audience won't notice.

"He also publicly proclaims that M-theory predicts *a great many universes were created out of nothing*. Not just the one we can observe. Existence can be derived from nonexistence by M-theory? That's interesting for a theory which has been described as *untestable* and *lacking predictive power*.

"Of course, prediction means before the fact, not after the fact. Stephen Hawking has a system that proves what he wants to establish, and he insists that the theory predicted it. His system predicts the winning horse, but after the race is over.

"Hawking's point has become clear. Science can do away with God because M-theory does a better job. It creates unlimited universes, rather than a single paltry one. God is redundant for the function of creation, and Hawking has arrogantly become the Donald Trump of science: *God, you're fired!*"

The defense attorney objects firmly. The judge instructs the prosecutor to stick to the facts.

"OK, but this interview doesn't bode well for Hawking's reputation; basing his cosmology on such shaky ground. His arguments, as well as M-theory, are not his original ideas. What we have from his testimony are mostly speculations and contradictions."

People may or may not like Hawking's interview, but after careful study of his book there is hardly any data from hisGrandDesign theory to back up his arguments. M-theory is really not evidence because it's untested. It's merely a novel hypothesis.

In any case, if Stephen Hawking represents where cosmology is at today then actual science has been abandoned. We're being fed a new religion called Scientism.

Did you think Hawking was trustworthy in this interview? I didn't get that feeling. Perhaps he's bluffing, or just being sardonic.

According to Wikipedia, Scientism has two equally pejorative meanings:

1. To indicate the improper usage of science, or scientific claims, in contexts where science might not apply, such as when a topic is perceived to be beyond the scope of scientific inquiry;

2. There is insufficient empirical evidence to justify scientific conclusions.

At this point it would be prudent to know what other renowned physicists think of M-theory. We don't have to blindly accept the word of one person just because he's a big name in physics.

Intelligent people don't accept things blindly, or out of awe of authority. They do their own research and come to their own conclusions by weighing the pros and cons from all sources, just as in a court of law. They're not dependent on one person's opinion.

In chapter two I present more research on M-theory. You will discover that several luminaries of physics and cosmology do not even consider it a theory.

2
Conflicting Testimony

After a two hour recess for lunch everyone is back in court. The prosecutor's next inquiry is to examine the opinions of other prominent cosmologists who are contemporaries of Stephen Hawking.

On a recent radio broadcast, *Unbelievable? with Justin Brierly*, the renowned physicist Sir Roger Penrose and Professor Alister McGrath discussed Hawking's book. The original broadcast was on Saturday, September 25, 2010.

The prosecution is ready to replay this discussion. The talk leads into the topic of whether Hawking's M-theory claim is good science.

Alister McGrath: "The first thing I want to emphasize is we have to be careful about the provisionality of this. In some of the debates on Hawking's book, you almost get the impression that science now knows the answer to this question and it's called M-theory.

"But it's not. It's just a staging post along the long road of science as we try to make sense of things. At the moment this looks quite hopeful, but it's clear that further work needs to be done. And the question is where will that take us in future?"

Roger Penrose: "Well, I think it's actually stronger than that. What is referred to as M-theory isn't even a theory. It's a collection of ideas, hopes, aspirations; it's not even a theory.

"And I think the book is a bit misleading in that respect. It gives you the impression that here is this new theory which is going to explain everything, and it's nothing of the sort. It's not even a theory and it certainly has no observational data."

Our prosecuting attorney stops the replay to offer a confirmation from renowned cosmologist Dr. Paul Davies who agrees with Penrose. In his book, *The Goldilocks Enigma*, he writes:

> "Nobody has yet written down the equations that govern the full M-theory, let alone solved them."[13]

"Well, this is certainly illuminating," affirms the prosecutor. "M-theory is actually just an idea. It is not based on data. Scientific theories, of course, require the support of evidence. At the very least they need a feasible way to gather evidence to either prove or falsify predictions.

"The scientific community may be cautiously excited about the potential of M-theory, due to its internal elegance and possible explanatory power, but the fact remains that it's not even theoretically testable. There is ongoing debate whether it can even be considered a theory.

"Most M-Theory adherents view it as a possible future scientific explanation. Columbia University researcher Brian Greene presents an overview of string theory, and its subsidiary M-

Theory, in his book *The Elegant Universe*. He clearly states the uncertain status of such theories and explains the implications.

> "No matter how compelling a picture string theory paints, if it does not accurately describe our universe, it will be no more relevant than an elaborate game of *Dungeons and Dragons*."

The prosecutor echoes the account of Greene. "Therefore, I contend that when evidence is eventually found to validate M-Theory, it will be a breath-taking achievement for modern science. Otherwise, it could turn out to be the most elaborate game of *Dungeons and Dragons* ever played by nerds. The coin is still in the air."

The next topic discussed on the show is whether Hawking's M-theory scheme shows that the universe created itself.

Justin Brierly: "This is one of the interesting aspects of what he appears to be saying, that M-theory shows that the universe can create itself, so to speak, out of nothing. Is that something which science can ever really tell us, or be definitive about?"

Roger Penrose: "It certainly is not doing it yet! I think the book suffers rather more strongly than many. I mean it's not an uncommon thing in popular descriptions of science to latch on to some idea, particularly things to do with string theory which have absolutely no support from observation. They're just nice ideas that people try to..."

JB: "But they will always remain theoretical..."

RP: "Always is too strong. It may well be they'll just be refuted, but even that is something which is very far from observational testability. They're hardly science."

Penrose continues to assert that science requires evidence. Later in the discussion another question introduces the multiverse idea: Is our fine-tuned universe simply one of many in a multiverse?

RP: "Multiverse means different things. There is this 10^{500} different M-theories, and these are different schemes in which you might have constants of nature being different. [10^{500} is 1 followed by 500 zeros. That's how many different universes are postulated, each with a unique set of laws]

"But not just that, all sorts of things are different; and the argument is, of all these different possibilities we live in one where life is possible. On most of them life is not possible. Now that's one type of scheme. They all kind of co-exist but we find ourselves in..."

JB: "...the one that is habitable for human life."

RP: "That's right, and that's what's called the Anthropic Argument, which has some justification. It's overused, I think. And this is the place where it's overused. It's an excuse for not having a good theory."

Pausing the broadcast, the prosecutor picks up on the point established by Penrose."Once again Roger Penrose establishes that M-Theory is not yet a theory. It's a collection of ideas and aspirations for a possible theory. Because it's untested it can be considered hypothetical at best, just as Ed Witten admitted.

"Again, Hawking uses M-theory in a way that is unwarranted, as an excuse for not having a good theory in the first place. This is the precise definition of Scientism.

"So it's clear that M-theory is not based on any observational evidence. First, we need data to support a theory. The absence of evidence means blind faith, according to Oxford University Professor Richard Dawkins."

Dawkins has declared religious faith to be:

> "...blind faith in the absence of evidence. I think it's a matter of belief without evidence. As a scientist and as an educator, I like the idea that we believe things because there's evidence."[14]

The prosecution continues. "M-theory is an untested and incompletely devised candidate hypothesis. In spite of this, Hawking invokes it as if it was truth, beyond doubt, that needs no further explanation or evidence at all. We have seen this with religious dogma, and now Hawking presents it in the name of Science. Once more, this is clearly Scientism."

Neil deGrasseTysoen, director of the Hayden Planetarium at the American Museum of Natural History in New York, is considered by some as the successor to Carl Sagan in popularizing science. Tyson notes:

> "There's a dimension to news reporting that I think not all journalists have the talent, frankly, to achieve, and that's to digest information, interpret it, and deliver it in such a way that people have a deeper understanding of what's going on."

Journalists can plead ignorance or unfamiliarity with the material – after all, they are not scientists. This does not apply to Stephen Hawking. In order to present a balanced case, the prosecution now wants to take a brief look at his career.

As a young man Hawking wanted to study physics under Britain's most notable physicist at the time, Sir Fred Hoyle of Cambridge University. But Hoyle was not accepting new students, so Hawking worked under the supervision of one of Hoyle's protégés, Dennis Sciama.

To be awarded a PhD at Cambridge, a student has to write an original thesis which is a substantial contribution to knowledge. This has to be accomplished within a three year period, so it's a heavy requirement.

Hawking couldn't find a suitable subject for his thesis. "I made a bad start at Cambridge," he explained in a TV documentary. "I had just been diagnosed with ALS, or motor neuron disease, and didn't know if I would live long enough to finish my doctorate, and I was having difficulty finding a problem for my thesis." [Hawking has struggled against Amyotrophic Lateral Sclerosis, a disease which left him almost completely paralyzed]

In an interview, Sciama recalled the situation.

Dennis Sciama: "In Stephen's case, his first couple of years was a bit slow because he was fallow at that time. He couldn't find a good problem and I couldn't find one for him."

With less than one year remaining, Hawking had still not come up with a worthy project. One day he attended a lecture by Roger Penrose, a young luminary in cosmology whose work on a problem connected with Einstein's theory led to an important discovery.

In his seminar, Penrose presented how large stars collapse to become black holes. Large stars collapse when they run out of fuel, but only a gigantic star can collapse into a black hole. The

29

entire star's matter is crushed into an infinitely dense point – a singularity.

At that time physicists believed this could only happen when a star was perfectly symmetrical. Then it would collapse uniformly in all directions. Could a star be absolutely symmetrical?

The work of Penrose established that if a star was large enough it could become a singularity whatever its shape. This was considered a very important contribution to the theory of relativity.

Hawking was inspired by Penrose and the work he was doing. One day he approached his supervisor with an idea for his thesis.

Dennis Sciama: "Hawking heard a seminar given by Penrose in which he announced his result. A little later Stephen said, 'We can adapt Roger's argument to the whole universe, in a certain sense that the universe is like a big star.'

"Of course, the universe is expanding, but if in your mind you reverse the sense of time, then the universe is collapsing. It's a bit like collapsing a very large star. Perhaps you can prove that in that collapse you can again achieve a singularity. It would mean that the Big Bang origin of the universe would have to be singular. So that would be a great discovery.

"So in his last year he proved his first singularity theorem for the universe on the basis of certain very reasonable assumptions; the Big Bang had to be singular."

In his own words, Hawking describes the outcome. "I was awarded my PhD by showing that Einstein's theory of relativity implied that the universe must have begun with a Big Bang. It couldn't have collapsed, bounced, then expanded again."

This became the Penrose Hawking singularity theorem. Penrose had developed the theorems to prove that massive stars collapse into a singularity.

Subsequently, Hawking was invited to the Vatican in 1975. The church was happy science had proven that the universe had a beginning, as stated in the biblical account, rather than the idea of the Steady State theory in which the universe always existed.

"In 1975, I was awarded a medal by the Pope for my part in proving the Big Bang theory."

This would not be Hawking's only visit to Rome.

"I went back to the Vatican in 1981 for a conference on cosmology, this time under a different Pope. He told us that it was fine to study the universe after the Big Bang but that we should not inquire into the Big Bang itself, because that was the moment of creation and the work of God."

Of course, Hawking wasn't a religious man so he didn't follow the advice of the Pope. The honor he received from the Vatican didn't change his view.

"If science and religion were now at one," he remarked, "perhaps they were still not quite seeing eye to eye."

Years later, Hawking related the story of meeting the Pope to an American audience. He added some humor to the tale.

"I was glad he didn't realize I had presented a paper at the conference suggesting how the universe began. I didn't fancy the thought of being handed over to the Inquisition, like Galileo."

Our prosecutor has a projector show the slide which Hawking presented at the conference showing himself in jail. The courtroom breaks out in laughter as the Pope becomes the butt of Hawking's joke.

Ironically, his work on the Big Bang theory moved science closer to religion. A beginning was the original premise of traditional creation stories. If there was a beginning then the conception that the universe always existed had to be revised. What science had been teaching for centuries was incorrect.

But that's not the end of the tale. Tomorrow the prosecutor will examine Hawking's book *The Grand Design* exactly as would be expected in a court of law. Will the book resolve contradictions we heard in his Larry King interview?

3
A Grand Fallacy

Many cosmologists today admit that even the best of their theories are simply hypotheses. Some of these hypotheses are unobservable and untestable. Few are based on objective data the way the scientific method prescribes and the way scientific empiricists work.

Several hypotheses are postulated as just being the most probable interpretation according to the specific model that a cosmologist favors. *The Grand Design*, a book by Stephen Hawking and Leonard Mlodinow, is a case in point.

Since the jury now has an idea of the modern creation story, the prosecuting attorney wants to determine if the book contributes anything substantial to our body of knowledge.

FIRST CHAPTER
Well, right on the first page, he notes, the authors take a dig at another discipline:

> "...philosophy is dead. Philosophy has not kept up with modern developments in science, particularly physics."

This condescending statement could be seen as an insult by many eminent philosophers at Cambridge University, who are actually colleagues of Hawking. Are we supposed to accept that these scholars haven't kept abreast of the world of science?

It's disappointing to see the book start on a negative note. Hawking's reputation as a brilliant physicist means he doesn't have to take a dig at others. Some physicists have a tendency to complain about philosophers and speak negatively about them, yet physicists apply philosophy which they fail to articulate properly, as we shall see.

For now, let's look at the definition of philosophy to clarify what Hawking considers to be dead.

Philosophy: the rational investigation of the truths and principles of being, knowledge, or conduct.

Clearly, a valid investigation of truths and principles of knowledge must be rational. Yet according to *The Grand Design*, rational investigation of truth by philosophers has died. Now only physicists can provide the light of knowledge? That's a challenge; one that the prosecution intends to pursue.

Next, a more detailed definition.

Philosophy: the academic discipline concerned with making explicit the nature and significance of ordinary and scientific beliefs, and investigating the intelligibility of concepts by means of rational argument concerning their presuppositions, implications, and interrelationships; in particular, the rational investigation of the nature and structure of reality (metaphysics), the resources and limits of knowledge (epistemology), the principles and import of moral judgment (ethics), and the relationship between language and reality (semantics).

The Grand Design begins with the denigration of philosophy, the discipline that eliminates nonsense from any discussions of reality. Why is that dead? Because, the book assures us, we can no longer use common sense for coming to conclusions about reality:

> "...common sense is based upon everyday experience, not upon the universe as it is revealed through the marvels of technologies such as those that allow us to gaze deep into the atom or back to the early universe."

From the first page the book establishes a negative approach, elevating its own discipline to a sublime level by discrediting the discipline of philosophy. Shall we consider the rigorous academic discipline of philosophy to be no more than mere everyday common sense?

Let's look at the definition of common sense.

Common sense: sound practical judgment; native intelligence.

We are asked to abandon native intelligence and sound judgment? Apparently, this will facilitate understanding the subject matter of Hawking's book. In order to accept the views of the book we can't rely on our own judgment and common sense. So on what basis do we accept the veracity of *The Grand Design*? It seems to be blind faith.

Right away, the opening pages of the book strike one as being familiar. Why it's just *The Emperor's New Clothes* in new dress.

35

This well-known tale by Hans Christian Andersen was first published in 1837.

> "An Emperor, who only cares about his appearance, hires two expert tailors. They promise him the finest suit of clothes made from a fabric so fine that it is invisible to people who are stupid, incompetent, or unfit for their position. The Emperor can't see the fabric himself, but he pretends that he can rather than appear unfit for his position, or stupid.
>
> "When the swindlers finish his new suit, they mime dressing the Emperor. Everyone in his court goes along with the pretense. When the monarch marches in procession before his subjects in his new clothes, every person continues on with the charade.
>
> Suddenly, a child in the crowd, too young to understand the desirability of keeping up the pretense, blurts out that the Emperor is wearing nothing at all. Then the cry is taken up by others.
>
> "The Emperor cringes, but holds himself up proudly and continues the procession under the illusion that anyone who can't see his clothes is either stupid or incompetent."

Dr. Hawking now drops a second bomb on his readers in his opening chapter. Everything in existence is "created out of nothing" according to the book's interpretation and presentation of M-theory. So now there's no need for the "intervention of a supernatural creator or god."

Fortunately, the fallacy of *everything in existence is created out of nothing* becomes crystal clear if we choose not to abandon intelligent judgment.

Nothing: no thing; not anything; naught; nonexistent.

Clearly, the science creation story promulgates that a singularity caused the Big Bang, so that was something that existed previous to creation. Hence, the universe arose from a singularity; it did not spring from nothing.

Secondly, nothingness has no properties because there is "no thing" existing that might have properties. Nothingness is a state devoid of anything that might even be a cause. So the idea that 'some thing' came from 'no thing' is a faulty conclusion. It's illogical and unscientific.

A more plausible and reasonable explanation might be that a singularity appeared from something which was imperceptible. Thus, it was misconstrued to come from nothing.

For example, if a singularity emerged from a higher dimensional reality that was undetectable, we would be wrong to conclude it sprang from nothing. That which is imperceptible does not imply its non-existence.

Another example: the mind of Stephen Hawking is unobservable. Should we conclude that his mind doesn't exist? It wouldn't be a philosophically sensible explanation, although his book assures us we can't rely on common sense.

Without sound practical judgment and native intelligence no person could ascertain if an argument was fallacious.

Fallacy: a deceptive, misleading, or false notion, or unsound argument.

The first major problem with Hawking's book is that he asks us to abandon reason and logic. He suggests this from the very first page. Only then can readers be led down the garden path to the book's conclusion. And those who do not accept this premise, are they incompetent to understand?

Isn't this the same explanation the swindlers gave in *The Emperor's New Clothes*? They spun cloth so fine that only the foolish couldn't see it.

In the absence of data, one may base arguments on any theory to prove a point. On the other hand, the Anthropic Principle – evidence that the universe is perfectly fine-tuned for life to arrive and thrive – is based on hard data.

Australian cosmologist Brandon Carter published a paper in the 1960s demonstrating that the laws of the universe appear to be a grand set-up to favor life. He called this bio-friendly fine-tuning the "Anthropic Principle." We also know it as the "Goldilocks Enigma."

The prosecution wants to make this case: "When everything in the universe is so finely tuned, it's difficult for cosmologists to escape the conclusion of prior advanced intelligence. Yet professor Hawking asks us to abandon normal intuitive reasoning. Why?

"Well, if the universe only *mimics* the activity of a bio-friendly intelligent system, then common sense reasoning would lead to a wrong conclusion. We ask the jury to consider: would a purposeless universe mimic a purposeful one?

"The observable evidence of bio-friendly laws that govern the universe, the mathematical laws that work, the laws of physics, the growing amount of evidence that the universe exists in clear exactitude for biological life, all of this evidence points to a more logical, reasonable, and plausible conclusion than *it just arose from nothing*."

Cosmology needs a good explanation to resolve this fine-tuning question. The modern creation story doesn't want to mimic the traditional story, after all.

The Grand Design attempts to respond to this challenge. The book declares that because our universe is fully suitable for biological life, with humans as the crown of creation, we can come to the following conclusion: "Although we are puny and insignificant on the scale of the cosmos, this makes us in a sense the lords of creation."

Yes you read that right. This statement is directly quoted from Hawking's book. But where is the correlation to the fine-tuning of the cosmos? Anyway, since Hawking brought it up I can offer a response.

I propose that the 100 percent death rate of every living person puts a nail in the coffin of the argument that humans are the lords of creation. If nature has 100 percent victory over us, in what way are we "the lords of creation?" Even so, Hawking will soon present an argument that we have no free will.

Professor Hawking explains the purpose of his book: to find an ultimate theory of the universe. "We now have a candidate for the ultimate theory of everything, if indeed one exists, called M-theory." Clearly, he's not sure that such a theory exists.

His book, however, intends to answer the why of existence rather than the how:

Why is there something rather than nothing?
Why do we exist?
Why this particular set of laws and not others?

This is strange because 'Why' is the realm of philosophy, and 'How' is the realm of physics.

Thus far, Hawking has never mentioned the name of Edward Witten who formulated M-theory. And he never does. But let's move on to see if his book can live up to its claim.

SECOND CHAPTER

Hawking opens his next chapter with that tired old idea of mocking other creation stories. This time it's Viking mythology. Yes, he wants to find a creation myth that appears implausible and use that to show the superiority of his own creation myth.

The two opening chapters have begun by denigrating different perspectives, including philosophy and common sense. Unfortunately, by adopting the approach that only physics is on the right track Hawking gives science the appearance of religious fundamentalism – only my view is true. If you don't accept my interpretation you must be ignorant or incompetent. It's a strategy. But is it a good strategy?

After a brief overview of the scientific method Hawking concludes: "Today most scientists would say a law of nature is a rule that is based upon an observed regularity and provides predictions that go beyond the immediate situations upon which it is based."

OK, we can accept that.

And, "...most laws of nature exist as part of a larger, interconnected system of laws."

Fine. No problem so far.

Then, "If nature is governed by laws, three questions arise."

1. What is the origin of the laws?
2. Are there exceptions, like miracles?
3. Is there only one set of possible laws?

These are good questions. They need good answers.

The traditional answer to the first question, according to Kepler, Galileo, Descartes, and Newton, was that God created the laws. Of course, we know from his opening chapter that Hawking believes the laws came into being from nothing, claiming the universe was caused by gravity and M-theory.

He clarifies the answer to his second question as, "...a principle that is important throughout this book. A scientific law is not a scientific law if it holds only when some supernatural being decides not to intervene."

It's clear by now that Stephen Hawking has a fixation about traditional creation stories. He allows this bias to color his work. I suggest he'd be better off dropping the negative tactic and adopt a positive approach. He could focus on what is wonderful about the scientific account of the creation story and avoid denigrating the other versions.

His entire treatise may be undermined by his attitude of damning the opposition, which is the way of politicians. Such a mentality turns people off politics and religion, and could also turn people away from Hawking's cosmology.

The answer to question three, "Is there only one set of possible laws?" turns out to be the thesis of Hawking's book. He believes there are infinite sets of possible laws. This will be discussed in a later chapter, he says, so we'll come back to it then.

The next idea offered in *The Grand Design* is that there's no free will. Nobody has the freedom to choose what they do or say. Yet, earlier he said we were "the lords of creation?"

> "Biology shows that biological processes are governed by the laws of physics and chemistry and therefore are as determined as the orbits of the planets. Recent experiments in neuroscience support the view that it is our physical brain, following the known laws of science, that determines our actions, and not some agency that exists outside those laws."

According to this, we don't control our brain because our brain controls us. Our choice to visit Paris or Rome or China is determined by known laws of science. Hawking never reveals which known laws of science determine what I eat, what I wear, and who I like or dislike. Thus, his premise remains in the realm of conjecture until he supplies the relevant evidence.

Professor Hawking states categorically that he's a scientific determinist. He believes the laws of science determine everything that happens in the past, present, and future. Again, this mentality mimics fundamentalist theists who say that God determines everything past, present, and future.

By denying free will and saying we are forced to act according to natural laws, Hawking affirms the existence of a higher power that has control over human beings.

The "higher power" in this case is the laws of physics, which he argues arose spontaneously from nothing. This argument mimics Biblical doctrine, so it's not his original idea. I'm not certain if he's being facetious or simply denigrating theology yet again.

He applies his determinist hypothesis to human nature and concludes there's no free will because everything is determined by scientific laws. Of course, the issue of the origin of the laws is conveniently brushed aside in favor of spontaneously arising from nothing at all. Again, we are in the realm of conjecture with a corresponding lack of scientific data.

He never reveals which laws he is talking about, yet he declares: "It is hard to imagine how free will can operate if our behavior is determined by physical law, so it seems that we are no more than biological machines and that free will is just an illusion."

A major problem with his determinism idea is this: We have to accept that there's no such thing as a criminal. People who choose to commit violent crimes are merely biological robots who are forced to act by the laws of physics. This would include Hitler, Stalin, serial killers, and suicide bombers.

I wonder if Hawking can live his philosophy. Would he call the police if burglars came into his home, or if a gang tried to molest his daughter? Would he acknowledge they are only acting under scientific laws, and should not be jailed?

Hawking's theory has zero value if the consequences don't apply to real world situations. Thus, his theory is just a mental construct. Nothing else.

Moreover, if laws of physics determine everything, then his own free will in writing his book is called into question. He can no longer be confident that what he writes is true, because it's coming from some unspecified law of physics acting upon him.

To live by his understanding, he should delete his name from his book and state, "written by an unspecified law of physics."

Ironically, the same physical law acts upon a theist who writes that God created the universe. The same law presents contradictory conclusions? Clearly this is nonsense.

What *does* make sense is that every person speaks according to one's own realization. In his own words, Hawking says he cannot imagine, "how free will can operate if our behavior is determined by physical law." This is the understanding of Stephen Hawking.

Of course, that doesn't exclude *others* from understanding how free will can operate despite physical law. Herein lies the fault of every skeptic. His own mindset is his binding truth. Whatever he can understand is his entire understanding – in spite of what others understand.

Hawking needs to provide some evidence for his conjecture. Otherwise his opinion is "blind faith in the absence of evidence" according to Richard Dawkins.

Let's challenge Hawking's argument with a thought experiment a la Einstein.

My body follows the laws of nature in all its functioning. In terms of chemical reactions, there's no law of nature it is ignoring. Yet, I have a will and intelligence over my body.

44

If I go for a swim, all the chemical reactions that govern breathing, heartbeat, and the flow of oxygen in my body change accordingly. These will change again if I'm reading in a library.

By the exercise of my free will I can manipulate my atomic chemistry and alter the functioning of my bodily machinery according to the specific purpose I want to accomplish. And yet, no physical law is ever broken or even modified.

Clearly, my mind does have a will over my body which functions according to the laws of nature. When we consider the accomplishment of human life, which is intelligent enough to examine its origins, it would be ludicrous to think that we have no free will.

Hawking should give more thought to this matter. He needs to answer these questions: Which physical law determined Mozart's or Beethoven's music? How about the music of Elvis Presley, or the Beatles?

Which physical law determined the religious fervor of Mother Teresa, or St. Francis of Assisi? Does the same physical law convince some people that God exists, and other people that God doesn't exist? If so, this physical law does not give predictable results.

Suddenly, the defense attorney rises from his seat to object to this line of reasoning.

"Your honor, some readers will conclude that Hawking is referring to ideas from biology, like adaptation/conditioning. They accept that the laws of physics govern free will because they determine the laws of chemistry, which determine the laws of biology, which determine the laws of psychology, which determine the laws of sociology.

"They would argue that opinions about beauty and genius are simply conventions from adaptation and conditioning which are explained by biology, which relies on biochemistry, which relies on physics.

"Mother Teresa's compassion might be seen as a favorable disposition to keep society from driving itself to extinction. People who believe in God could be dismissed as simply naïve, or maybe less-evolved people who will eventually die out along with their creation stories. Dishonest people are biologically, genetically, or circumstantially driven to such behavior. Ultimately, the laws of physics prevail."

The prosecution counters with this argument. "Nevertheless, your honor, people who posit such arguments insist that criminals who molest their children be dealt justice. Their theories and opinions are creations of their own views, which in practice are not useful for the public good."

The court concurs. The problem, of course, is that no scientist could ever provide data to substantiate the non-existence of free will, due to the insurmountable complexity involved in determining action. So it remains in the jurisdiction of speculation.

When we discuss science there is a need to substantiate hypotheses with data. Without evidence we have conjecture, or its close cousin, blind faith, both of which live in the "no data" region of existence.

If Hawking is certain that some physical law determines whether people become honest rather than dishonest, or vice versa, and that they are driven to such behavior, then he needs to specify which physical law he has pinpointed to relieve our convicted felons of their guilty status. Unfortunately for the convicts, his *Grand Design* does not specify any such law.

Hawking is losing credibility. He has a particular mindset that he projects onto the picture of reality, and thereby constructs a "theory" to substantiate what he *wants* to be true. But with no data we're left with blind faith, and that's not science. Maybe it's reverse engineering in reverse.

The prosecutor decides to plod on in order to identify the answers to the three 'why' questions.

4
Same Wine, Different Bottle

Day three in the courtroom.

The prosecution continues examining the logic and science of Hawking's Grand Design theory of the origin of our universe. The book is beginning to give the impression that he has lifted religious doctrine and applied it to cosmology. Perhaps, he assumes that if 'no data' is acceptable for religion then it should be acceptable for science, too.

Clearly, however, it's evidence that separates science from religion.

THIRD CHAPTER
At this point, Professor Hawking wants to explain the true meaning of objective reality. What we think is reality, he says, may only be the simulated reality of websites such as *Second Life*.

He admits that his idea comes from the science fiction film *The Matrix*, "in which the human race is unknowingly living in a simulated virtual reality created by intelligent computers to keep them pacified and content while the computers suck their bio-electrical energy."

His book offers various concepts and models of reality, but too many are science fantasy. These are not innovative ideas. They can be found in science fiction, Zen Buddhism, and yoga philosophy. Most yoga adherents understand the concepts of Illusion, *maya*; and Truth, *tattva*.

This chapter is just the same old wine in a different bottle. There is no recognition that these are not scientific ideas, just philosophical ones which other traditions have already sorted out.

To the informed reader, the modern jargon unsuccessfully disguises these ideas as fresh. Where are the new ideas? We don't want a rehash or another *Matrix* sequel.

Eastern philosophy offers a positive contribution to the realm of thought. We may live in an illusory objective world where everything is temporary, but a true reality does exist and can be experienced.

The very concept of illusion relies on the fact of an actual reality. For example, mistaking a length of rope for a real snake implies that snakes exist. The goal of life, according to Eastern philosophy, is to realize true reality through the yoga process. That's how we escape illusion.

When knowledge appears, ignorance disappears. When the Sun appears, darkness vanishes. A rope can only be mistaken for a snake in the dark. Reality is a different nature than an experience that props up the illusion.

In contrast to the philosophy of the East, Hawking states that we can't know reality. Does this mean we remain in illusion forever? Then it must mean Professor Hawking is also under illusion. So why should we take him seriously? Again, this is the rationale of the skeptic. If he can't know reality, neither can anyone else.

In spite of this, he will now provide a model to help us understand the unknowable. First, he explains the four qualities of a good model:

1. A good model is elegant
2. It agrees with and explains all observations
3. It has few arbitrary or adjustable elements
4. It can make predictions about future observations that disprove the model if they are falsified.

Then he offers a perspective which is important to understand his book:

> "There is no picture- or theory-independent concept of reality. Instead we will adopt a view that we will call model-dependent realism: the idea that a physical theory or world picture is a model (generally of a mathematical nature) and a set of rules that connect the elements of the model to observations."

A particular understanding of reality is based on some particular model, he says. But reality is independent of any model thus far designed by physicists because they are constantly adjusting their models.

Hawking assures us that model-dependent realism "can provide a framework to discuss questions such as: If the world was created a finite time ago, what happened before that?" Interesting. But does it relate to the real world in which we all live?

Apparently not. "According to model-dependent realism it is pointless to ask whether a model is real," he says, "only whether it agrees with observation."

Yet, earlier in his book, Hawking discussed the model of Ptolemy and dismissed it, although it *did* agree with observation. Ptolemy's model was rejected even though Hawking assured us that "it is pointless to ask whether a model is real, only whether it agrees with observation." Evidently, the old professor doesn't follow his own advice.

However, he does admit that it is not "clear yet whether a model in which time continued back beyond the Big Bang would be better at explaining present observations because it seems the laws of the evolution of the universe may break down at the Big Bang."

Although he posits a framework to discuss what happened before creation, he concludes that even if we had such a model it might not help us. Why? Because the laws that govern the evolution of the universe apparently break down at the point of creation. He's not certain, but Einstein's equations "may break down at the Big Bang." That's his informed opinion.

Let's take a second look at Hawking's PhD thesis. Perhaps you remember that he said, "I was awarded my PhD by showing that Einstein's theory of relativity implied that the universe must have begun with a Big Bang."

By going backwards in time, he posited that the universe would resemble the collapsing star of Roger Penrose's theory. Penrose developed the singularity theorems that Hawking borrowed to prove that by going back to the Big Bang we arrive at a singularity.

But now in the 21st century, going back in time means physical laws break down at the Big Bang. Mimicking Penrose's theory is no longer a viable concept to understand the "bang" in the Big Bang. It means his PhD thesis was a conjecture that looked true at the time, but no longer looks true. Should he renounce his PhD title now?It's no longer a substantial contribution to knowledge.

We are left with this issue: if physical laws break down as we approach the Big Bang, how do we know they existed prior to that? The stuff from the Big Bang expanded at extremely high velocity far greater than the speed of light. Thus, the universe grew from tinier than an atom to an unknown enormous size in a fraction of a second.

The hypothesis of instant expansion (cosmic inflation) defies all known laws of physics. If we accept that cosmic inflation defied all known laws of physics, then what caused inflation?

If inflation was not caused by any known laws of physics, does it imply that the laws at creation were supernatural in origin? Or, did the laws simply arise after creation and not before? If so, what is that mechanism that manufactures universal laws?

If we were hoping for clarity in regards to reality, it looks like another speculative walk down the garden path. But Hawking will stick to his guns because "physicists are indeed tenacious in their attempts to rescue theories they admire..."

Is "rescuing theories they admire" the job description of physicists? The word "rescue" means saving someone or something from peril. The imperiled theory Hawking intends to rescue appears to be the modern creation account.

He cites former models of the universe that fell by the wayside, "such as the four-element theory, the Ptolemaic model, the phlogiston theory, the Big Bang theory, and so on. With each theory or model, our concepts of reality and of the fundamental constituents of the universe have changed."

Clearly, by including the Big Bang on the list of redundant theories, Hawking admits that it is no longer viable. He assumes that his Grand Design model will render the Big Bang model obsolete. No lack of bravado on his part.

We may note that reality remains the same, as it was before so it is now. Only our ability to observe reality has changed, and thus our conceptions.

Next, Hawking claims the following:

> "In the 1920s, most physicists believed that the universe was static. Then, in 1929, Edwin Hubble published his observations showing that the universe is expanding."

Did Hubble really believe the universe was expanding? Not according to the book he published in 1937, as well as numerous papers after 1929. I deal with this extensively in Chapter Eight of my book, *Cosmology on Trial*.

Regarding the laws that govern the universe, Hawking concludes: "There seems to be no single mathematical model or theory that can describe every aspect of the universe. Instead, as mentioned in the opening chapter, there seems to be the network of theories called M-theory. Each theory in the M-theory network is good at describing phenomena within a certain range."

So it seems the M-theory model will challenge the Big Bang model. Hawking believes that Ed Witten's M-theory is the answer, although Witten's name is never mentioned. Instead, the model is called the Grand Design.

Hawking is unaware we have researched M-theory and discovered it's merely a set of ideas, hopes, and aspirations. It can't be called a network of theories, as defined by genuine scientific methodology. If we accept the statements of reputed physicists themselves, it's more like a herd of hopeful hypotheses.

Ironically, even a network of theories is inelegant. It's similar to Ptolemy's epicycles. This defeats Hawking's first premise of a good model. Of course, if we accept his framework he can lead us to his own conclusions. But if his framework is faulty, or if new discoveries demonstrate his model to be inaccurate, then we end up with ignorance, not knowledge.

Hawking is full of hope that the grand design model will not fall by the wayside. At this point, however, the prosecution is not as hopeful.

For the rest of his book Hawking uses his own brand of philosophy, not science, to support model-dependent realism. Again, I want to emphasize that reality pre-exists any model he can imagine. He is desperately trying to shoehorn reality to fit his ideas about M-theory.

This mentality of gambling with "knowledge" which can lapse into ignorance within a few decades is of little benefit to humanity.

Hawking's logic in this chapter is not sound, and that doesn't bode well for the rest of his book.

5
Quantum Physics & Consciousness

So far we have encountered contradictions and speculations that lack evidence in the Grand Design theory. Is there more of the same?

The prosecuting attorney questions whether we will finally get some genuine physics, and even observational data to substantiate the conjecture.

FOURTH CHAPTER

The fourth chapter of Hawking's book begins by stating that Newtonian physics was inadequate to describe how nature behaves at the atomic and subatomic level. This led to the emergence of quantum mechanics. For Hawking, the fact that atomic particles behave differently than described by Newton's laws means we can no longer rely on intuitive logic.

His justification is that reality "is a picture in which many concepts fundamental to our intuitive understanding of reality no longer have meaning."

He remarks that the components of normal objects, atoms and molecules, obey quantum physics. But composite objects like tables and chairs behave according to Newton's laws. Newton's laws match the view of reality we experience in everyday life. Atoms and molecules reveal a different version of reality. So again, Hawking asks us to abandon common sense.

By definition, common sense does not apply to objects which are beyond our purview, like atoms. It's only useful for the everyday version of reality. So Hawking's idea to abandon common sense for a different version of reality where it does not apply is not coherent.

The traditional creation story also defined God as a different version of reality by applying the concept of a supernatural entity. This is not what Hawking is talking about.

"Physicists are still working to figure out the details of how Newton's laws emerge from the quantum domain," he admits. "What we do know is that the components of all objects obey the laws of quantum physics, and the Newtonian laws are a good approximation for describing the way macroscopic objects made of those quantum components behave."

At this point several features of quantum physics need to be explained:

1. The wave/particle duality. Photons of light behave both like a wave and like point particles.

2. The Heisenberg Uncertainty Principle, formulated in 1926 by Werner Heisenberg, explains that our ability to simultaneously measure the velocity and the position of a particle is severely limited.

3. The Uncertainty Principle states: "if you multiply the uncertainty in the position of a particle by the uncertainty in its momentum (its mass times its velocity) the result can never be smaller than a certain fixed quantity called Planck's constant."

4. The more precisely we measure the position of a particle the less accurate is its velocity, and vice versa.

If we double the uncertainty for speed, then we will need to halve the uncertainty for position. Hawking states what this means: "given the initial state of a system, nature determines its future state through a process that is fundamentally uncertain."

The above explanations might not be readily comprehensible, but you can feel secure in the knowledge that quantum theory also bothered Einstein. That's why he became critical of it.

Richard Feynman, another great scientist of the 20th century, wrote, "I think I can safely say that nobody understands quantum mechanics." Are physicists any closer to that understanding today?

Scientific theory has two components: the mathematical core, and the physical interpretation of the mathematical equations. Quantum mathematics is substantiated, but the physical interpretation is so obscure that Einstein and Feynman had to speak up.

Hawking and Mlodinow, however, assure us that they understand quantum mechanics. Quantum physics might seem to sabotage the laws of nature, but Hawking affirms that nature determines the *probabilities* of sub-atomic events rather than choosing a definite outcome. He then extends the indeterminate action of sub-atomic particles to the universal scale and makes a claim for a new universal determinism.

This idea "leads us to accept a new form of determinism," he writes. "Given the state of a system at some time, the laws of nature determine the *probabilities* of various futures and pasts rather than determining the past and future with certainty."

Of course, quantum physics only deals with the sub-atomic level. On the level of reality of everyday life, there is the thoroughly predictable Newtonian physics based on laws that govern familiar matter. Hawking has already admitted the difference but now passes the uncertainty of the sub-atomic level onto the scale to which it does not belong – familiar matter which we can perceive. Does he see the contradiction?

Apparently not, because he asserts that quantum theory should be interpreted in such a way that the past is no longer fixed history as we generally believe. Hawking's book acknowledges that some people may find this idea offensive. Others may find it ridiculous.

Can this interpretation be verified? Of course, it has never been tested. Again, the reader is asked to abandon sound judgment. Now, he claims, "Scientists must accept theories that agree with experiment, not their own preconceived notions."

Good point. Experiment must verify theory. Can Stephen Hawking live up to his claim? Has any macro-system of matter, from a marble to a galaxy, ever been proven experimentally to behave as sub-atomic particles do? Hawking does not provide any evidence.

Moreover, the Uncertainty Principle of quantum physics presents a major interpretive conflict. It proposes that the movement of particles in sub-atomic reality behave in a manner that is uncertain. However, this uncertainty could be generated by a certain lack of knowledge.

If we can never accurately measure both velocity and position of a particle simultaneously, it might mean that the movements of the particles are indeterminate. It could also mean that the observer was unable to determine their behavior due to incomplete knowledge.

Einstein and other physicists postulated the "hidden variable theory" to recognize incomplete knowledge. They felt there was a more fundamental theory hidden beneath quantum mechanics whereby the entire system would reveal a deeper reality which could predict outcomes with certainty. They proposed that the "Copenhagen interpretation" of quantum theory by Neils Bohr was an incomplete description of reality.

Quantum physicists were quite certain about the nature of uncertainty, so the hidden variable idea lost popularity over time. But now theorists are exploring ways to bring back the idea with modifications because it's foolish to assume that we know it all and there is nothing left to be discovered.

To reduce the burden of uncertainty on sub-atomic particles, Hawking assures us that if an experiment is repeated enough times the data will represent the probability of a particle's position and velocity. This would make it scientifically valid. He accepts Richard Feynman's theoretical sum-over-histories model, and uses it as a practical paradigm to try and solve the indeterminacy problem.

According to Feynman, a system takes all possible paths which sum up to a probability amplitude. But there is no determinism in the sense that there is no specific answer at the end, just probabilities.

The result is we are not left with a certain event, only a theoretical probability – an approximation of outcomes – because no one can know all the possibilities. Nor can one even assume that all the possibilities are observable. That's why the word "probability" is appropriate and not the word "certainty."

Generally, books and papers on quantum theory discuss the various interpretations and observable effects, but also state that science does not know the correct answer at present. Stephen Hawking misrepresents the *status quo* of quantum physics by presenting that science has moved forward in its understanding of quantum mechanics and an established accepted understanding does exist.

Basing a grand scientific conclusion on a particular explanation without acknowledging there are deficiencies in the argument, and there are other valid alternative explanations, misleads people to accept that one particular view is actually correct.

Furthermore, interpretation is not the same as the knowledge, or proof, of a thing. Cosmologist Sean Carroll describes the problem of interpretation in quantum mechanics as the "greatest embarrassment" to modern physics. He says he's embarrassed because physics is looking less like science and more like philosophy (where everybody gets to put in their two cents worth).

Until there is data to distinguish between interpretations, each version is like a beauty pageant contestant striving to emerge victorious. Needless to say, that is not science.

Theorists, however, can take advantage of this new face of science where interpretation comes forward and proof is left on the back burner. It means they can publish hundreds of papers and nobody can prove them wrong.

Quantum mechanics is a wonderful theory. It can explain things that were unexplainable previously. However, Sir Roger Penrose points out a fallacy in the September 2009 issue of *Discover* magazine:

> "When you accept the weirdness of quantum mechanics [in the macro world] you have to give up the idea of space-time as we know it from Einstein. The greatest weirdness here is that it doesn't make sense. If you follow the rules, you come up with something that just isn't right."

The quantum description of reality appears contrary to the reality we experience. Quantum theory postulates that a single object can be in many places simultaneously. Penrose takes issue with this idea in his *Discover* interview:

> "It doesn't make any sense, and there is a simple reason. You see, the mathematics of quantum mechanics has two parts to it. One is the evolution of a quantum system, which is described extremely precisely and accurately by the Schrödinger equation. That equation tells you this: If you know what the state of the system is now, you can calculate what it will be doing 10 minutes from now.

"However, there is the second part of quantum mechanics—the thing that happens when you want to make a measurement. Instead of getting a single answer, you use the equation to work out the probabilities of certain outcomes. The results don't say, "This is what the world is doing." Instead, they just describe the probability of its doing any one thing. The equation should describe the world in a completely deterministic way, but it doesn't."

Erwin Schrödinger is regarded as a genius. Then, accordingly, did he recognize the conflict his equation created? Penrose has no doubt about this:

"Schrödinger was as aware of this as anybody. He talks about his hypothetical cat and says, more or less, 'Okay, if you believe what my equation says, you must believe that this cat is dead and alive at the same time.' He says, 'That's obviously nonsense, because it's not like that. Therefore, my equation can't be right for a cat. So there must be some other factor involved.'"

Clearly, if it's not right for the cat, then it can't be right for other macro objects. Penrose confirms that this is what Schrödinger is pointing out.

To accept that the cat is dead and alive at the same time is a reality that no one has ever experienced. It's a superimposition of two contradictory conditions existing simultaneously. Theoretical physicists justify this bizarre cat analogy as a reality we must acknowledge in order to rationalize that our consciousness can take other paths without our knowing it.

Can we verify what can't be experienced, or even known? Yet this conception is motivating many ideas in theoretical physics towards the many worlds interpretation of reality. The interpretation posits that all probabilities exist somewhere in some parallel universe.

Penrose says, "You're led to a completely crazy point of view. You're led into this many worlds stuff, which has no relationship to what we actually perceive."

Stephen Hawking should honestly admit that his theory is really just an interpretation until it's verified by data. He also fails to mention the mathematical problems involved in his interpretation. By not being completely candid, he leads his readers to accept what Einstein and Feynman did not accept.

CONSCIOUSNESS

Richard Feynman wrote that the famed double-slit experiment "contains all the mystery of quantum mechanics."[15]

[youtube.com/watch?v=DfPeprQ7oGc]

When particles pass through two slits, they form a pattern of two lines on a back screen as a result. When waves pass through the same two slits, they form a series of lines of different intensities on the back screen called an interference, or diffraction, pattern. However, when electrons pass through the double slit, they imitate the trajectory of waves.

How could particles of matter create an interference pattern like waves? Physicists decided to place a measuring device on one slit to detect what happens when each electron passes through it. Then the mysterious nature of the subatomic world was revealed. Suddenly, the electrons acted as particles and formed two neat lines. Did they recognize they were being observed?

Feynman came to the amazing conclusion that the particles in the double-slit experiment had some sort of information. As soon as an observer tried to track the particles, they behaved differently. The act of measuring collapsed the wave function.

Light had always been accepted as a wave, but Einstein said that light was a particle, a photon, and his theory supported that conclusion. Yet when light goes through the double slit it forms a diffraction pattern of many bands, like a wave. So how can it be a particle?

Once again, physicists set up a measuring device to find out what's going on. They fired photons one at a time through the double slit and this time the result was two lines, as would be expected from a particle. Again, the act of measuring collapsed the wave function. Were the detectors interfering?

Heisenburg explained that the experimental apparatus itself alters the sub-atomic particles' trajectory because the particles we are trying to observe are of the same scale as the photons we're using to observe it. In other words, to observe something we must bounce photons off it. But the photons disturb the particles because they are the same size. So there is no way to observe sub-atomic particles without altering their trajectories.

The double-slit experiment was repeated with the detectors still in place but with no data being recorded. The detectors were detecting but not collecting the data. However, this time the photons made a diffraction pattern.

To recap: when physicists collect data, photons act like a particle. If they don't collect data, photons act like a wave. In both cases detectors are present. Does the conscious act of collecting data determine whether light behaves like a wave or a particle?

Erwin Schrödinger posited that since the photons don't break up, they must simply exist as a probability distribution, rather than a particle, before any measurement is made. As waves of probability they interfere with themselves as they pass through the slits to cause a diffraction pattern on a back screen. When no measurement is made, there is just probability. If a detector is utilized at the slits, the probability wave function collapses to a physical particle.

Clearly, in the double-slit experiment there seems to be two laws at work: one that applies when you're detecting, and another that applies when you're not detecting. This attaches some importance to the awareness of the observed as well as the observer. In other words, reality seems to be a product of some form of consciousness, as noted by Nobel Laureate Eugene Wigner.

> "It will remain remarkable, in whatever way our future concepts may develop, that the very study of the external world led to the scientific conclusion that the content of the consciousness is the ultimate universal reality."

Max Planck echoed the same realization as Wigner: "Science cannot solve the ultimate mystery of nature because, in the last analysis, we ourselves are a part of the mystery that we are trying to solve."

Even more amazing is the mysterious phenomenon of entanglement of particles.

[youtube.com/watch?v=1BfJ06plOTs]

When two atoms are entangled with one another they share a peculiar connection that is instantaneous and unaffected by distance. Experiments establish that even over a vast stellar expanse, the atoms remain connected. Whatever happens to one instantaneously affects the other. This behavior is so synchronized, like a choreographed dance, that it stretches the concept of communication. They are so connected, that the word "entanglement" is used. Einstein disliked the concept, calling it "spooky action at a distance."

Of course, if particles can transfer information we must ask, how do they acquire it, how do they transmit it, and where does the information come from? Such knowledge implies some form of intelligence, which means some form of consciousness.

Schrödinger affirmed that entanglement, which is the concept of connectivity, is factually the basic property of quantum mechanics. Quantum mechanics, then, is the display of waves of potentiality. It's the field of pure potentiality, of abstract potential existence. Connectivity among all things is a basic constituent of the fabric of reality.

Feynman's interpretation avoided the conclusion reached by Wigner and Planck. Sidestepping the taboo subject of consciousness, he introduced a different perspective: that the particles take every possible path rather than taking one definite path. Moreover, they seem to take every possible path simultaneously. But this offered no explanation why particles behave differently when measured.

It's clear that humans can't know every possible outcome, yet it appears that particles can, according to Feynman. Thus, we again come to Wigner's conclusion of reality being a product of some form of consciousness.

Feynman undermines Hawking's theory that the universe simply popped out of nothing if it contains all the symptoms of intelligence, including particles with prior knowledge and finely-tuned laws of mathematics and physics.

Alluding to Feynman's explanation, Professor Hawking writes: "There are an infinite number of paths, which makes the mathematics a bit complicated, but it works."

The Grand Design theory must be his personal conviction that it works, but the concept of infinity does more than make the math just "a bit complicated." Every mathematician knows, and Hawking also knows, that we cannot do mathematics with infinity.

A true infinite has no beginning and no ending. Mathematicians may say there are infinite numbers in the set from 2 to 8. But that infinite has bounds, a beginning 2 and an end 8. Moreover, the numbers 1 and 9 are not included in this infinite set. But there is no question of doing mathematics with an actual infinite.

Physics studies physical things, which have a beginning and an end. The Big Bang concept establishes that the universe is physical because it had a beginning. And the particles of quantum physics had their beginning also. An actual infinite, on the other hand, means something beyond physical nature, a super-nature without beginning or end.

By the following four tenets physics and cosmology paint themselves into a corner in which the modern creation story seems trapped:

1. Physical nature is all that exists
2. Supernatural events do not exist
3. M-theory posits infinite natures
4. These natures can never be supernatural

Modern science doesn't allow anything supernatural to enter its story because that would grant some validity to the traditional story.

Still, in contrast to premises 2 and 4 above, the existence of dimensions beyond our present understanding is a possibility according to M-theory. There's no data to prove it now but like dark energy, it could be discovered soon.

6
A Theory of Everything?

Day five in the courtroom. The prosecution is on a roll while the defense attorney has not been able to present a difference of opinion to the historical arguments.

The title of Hawking's fifth chapter is "The Theory of Everything." Really? Not just physics stuff, but everything else like cures for disease, old age, and death? Surely these are included in everything.

His chapter begins with an Einstein quote: "The most incomprehensible thing about the universe is that it is comprehensible."

Professor Hawking accepts this conclusion because the universe is governed by "scientific laws." With all due respect, the old professor is not expressing himself properly. The entire cosmos is governed by laws of nature, which science is gradually discovering.

Hawking's presentation is misleading. Almost a dozen theories of electromagnetism existed up to the 1860s, for example , yet all

71

of them were flawed. Then, James Maxwell discovered the mathematics that explained the relationship between light, electricity, and magnetism. Hawking writes, "Maxwell had unified electricity and magnetism into one unified force."

Factually, Maxwell did no such thing. He was able to demonstrate that it was *already* one unified force. Maxwell's "laws" already existed and governed universal forces billions of years before Maxwell was born.

The fact that various branches of science are discovering natural laws hardly makes them the laws of science. Nature's laws existed for millennia before science claimed credit. Moreover, if our understanding of the laws changes over time, nature does not oblige us with different behavior accordingly. Once again, the old professor isn't expressing himself properly.

Nevertheless, in 1687 Isaac Newton presented the law of gravity as a mathematical formulation. Every object in the universe attracts every other object with a force proportional to its mass; the familiar Inverse Square law of gravity.

Subsequently, scientists discovered that various phenomena of the universe could be modeled if the mathematical machinery was understood. They began discovering mathematical formulations like explorers discovering new lands, and claiming proprietorship over the same.

The question remains unanswered: How did the laws of physics and mathematics first come into being so that scientists could discover their existence?

Unfortunately, Hawking skirts the issue of the origin of natural laws. He sweeps it under the rug and pretends that the house of physics is clean. But he needs to explain why these universal laws exist, because he has promised this from his first chapter.

At this juncture in his book, Hawking introduces string theory.

> "According to string theory, particles are not points, but patterns of vibration that have length but no height or width – like infinitely thin pieces of string. String theories also lead to infinities, but it is believed that in the right version they will all cancel out. They have another unusual feature: They are consistent only if space-time has ten dimensions, instead of the usual four."

The strings are one-dimensional, but the math only works if there are ten dimensions? Of course we only observe three plus time, which makes four dimensions – Einstein's space-time.

Physicists are leading us from the unobservable to the inconceivable, and now to the unprovable. If that doesn't complicate our picture of "reality" enough, we find the mathematics points to not one, but five possible string theories.

"String theorists are now convinced that the five different string theories and super-gravity are just different approximations to a more fundamental theory, each valid in different situations," explains Professor Hawking.

The more fundamental theory he refers to is M-theory, which was mentioned earlier: "People are still trying to decipher the nature of M-theory, but that may not be possible."

It may not be possible to decipher M-theory, but Hawking will give it a try. This is the first disclaimer for his idea of a grand unified theory of everything using M-theory.

Another disclaimer: "It could be that the physicist's traditional expectation of a single theory of nature is untenable, and there exists no single formulation."

The third disclaimer: "It might be that to describe the universe, we have to employ different theories in different situations."

How then is it one unified theory as claimed? It seems Hawking's assertions follow the quantum laws of uncertainty. We can name these disclaimers: the Hawking Uncertainty Principle of M-theory.

> "Each theory may have its own version of reality, but according to model-dependent realism, that is acceptable so long as the theories agree in their predictions, that is, whenever they can both be applied."

One grand unified theory (GUT) may not be possible. Nonetheless, Hawking suggests utilizing different theories for different aspects of the universe and still have a unified theory. Of course, it won't be as elegant as one GUT. In fact, far less elegant than Ptolemy's model which was rejected even though it gave accurate approximations of orbital motion.

Hawking is sold on M-theory and promotes it vigorously. As he stated earlier, "physicists are indeed tenacious in their attempts to rescue theories they admire..."

In his book, Hawking acts like a lawyer who represents M-theory just like he did on the Larry King Show. Apparently, it does need defending because reputable physicists affirm that it is untestable, thus stretching the boundaries of good science.

Still, Hawking would have us believe that the theory has now become law. In his own words:

> "The laws of M-theory therefore allow for *different universes* with different apparent laws, depending on how the internal space is curled. M-theory has solutions that allow for many different internal spaces, perhaps as many as 10^{500}, which means it allows for 10^{500} different universes with its own laws." [his emphasis]

We have already ascertained that M-theory is simply a collection of ideas, yet Hawking presents these ideas as "the laws of M-theory." But if the goal is to unify all natural laws into one GUT, M-theory is a giant leap in the wrong direction when it predicts 10^{500} different possible worlds, all with different laws, only one of which would be our universe.

Yet Hawking's book makes an appeal for this multiverse concept, a conception which multiplies the probability that the fine-tuning of the cosmos came about by chance. Let's examine the concept.

We know the early conditions after the Big Bang explosion were finely calibrated for intelligent life to arise. If conditions are adjusted by the tiniest amount, life would be impossible. Hawking doesn't deny fine-tuning. He even qualifies it as almost miraculous. The universe does look like it's been exquisitely designed to produce intelligent life. Hawking won't dispute that. But after admitting it looks like that, he explains it via the multiverse idea.

Does the unlimited universes idea account for fine-tuning? With unlimited roulette wheels the chance that my number will come up on at least one of them is greatly enhanced. With unlimited universes, each having a unique set of laws, the chance that one universe is fine-tuned for life, or bio-friendly, is a reasonable expectation from the probability laws. And that universe might just as well be ours.

However, this is simply a trivial solution, not a rigorous proof, because we cannot prove the existence of these other worlds. What to speak of examining their laws to see if they are bio-friendly. Such conjecture, however, has a respectable name: it's called the Weak Anthropic Principle.

The multiverse conception postulates a set of universes rather than one universe (uni = one, multi = many). It came about by physicists speculating that instead of our one universe there might be various different cosmic regions resembling a patchwork quilt of universes with differing physical laws and properties.

From this hypothesis, Leonard Susskind of Stanford University tried to resolve the Anthropic Principle, the situation of our universe being fine-tuned for life, in his book *The Cosmic Landscape*. He posited that life obviously couldn't survive in a cosmic region where life was impossible, so it's not astonishing that life thrives in our corner of the multiverse. Thus, he created the weak anthropic principle.

Susskind's conjecture was happily greeted by scientists who weren't happy that the fine-tuning of the universe evoked the dreadful God concept. Professor Paul Davies described it like this:

> "...they seized on the multiverse theory as a neat explanation for the uncanny bio-friendliness of the universe."

Apparently, Hawking anticipated an outrage for his use of the multiverse to explain fine-tuning. Thus, we find another disclaimer in his seventh chapter. The "multiverse idea is not a notion invented to account for the miracle of fine-tuning. It is a consequence of the no-boundary condition as well as many other theories of modern cosmology."

The "no-boundary condition" refers to the assumption that the universe is infinite which would allow for a possible multiverse to exist. But then we find an inconsistency in Hawking's same chapter. Although he claims the multiverse idea was not invented to account for the miracle of fine-tuning, every physicist uses it to account for the miracle of fine-tuning.

He writes: "in the same way that the environmental coincidences of our solar system were rendered unremarkable by the realization that billions of such systems exist, the fine-tunings in the laws of nature can be explained by the existence of multiple universes."

There is no data to support "the existence of multiple universes." It's an unproven hypothesis that Dr. Hawking arbitrarily elevates to genuine existence. Does an unproven hypothesis validate multiple universes? Obviously not. Yet Hawking does prop up a so-called "realization" to promote his grand design model in his next example.

> "Just as Darwin and Wallace explained how the apparently miraculous design of living forms could appear without intervention by a supreme being, the

multiverse concept can explain the fine-tuning of physical law without the need for a benevolent creator who made the universe for our benefit."

So, the multiverse concept *is* promoted as a notion to account for fine-tuning. Hawking's later statements contradict his earlier ones. It's one contradiction after another. This sleight-of-hand makes his theory look like a swindle.

Next, we will examine Hawking's claims in more detail. Does the multiverse concept factually explain the fine-tuning of the universe? And since he brought it up, does it rule out a benevolent creator in the process?

If M-theory predicts unlimited possibilities it means anything and everything is possible. Therefore, if a theologian claims that God created a universe where Jesus Christ died for people's sins, science can no longer deny this possibility. M-theory allows for it in a scenario with infinite universes and infinite possibilities. Hawking's model can't escape the logic that the greatest likelihood of this event was in our own universe.

At this juncture, the defense sees an opening and catches the attention of the judge to make his argument.

"Your honor, my colleague may argue that the probabilities proposed by M-theory refer to different laws of nature as well as different parameters in the laws, i.e. different particle masses, charges, etc. However, it does not predict universes where a supernatural force may be present."

The judge defers to the prosecution for a response.

"Well, the argument by my esteemed colleague that M-theory can predict whether or not a force is supernatural, is an arbitrary assumption on his part. It is not substantiated by experimentation, observation, or any method accepted as scientific. The defining statement of M-theory, infinite universes with infinite possibilities, does not exclude any particular outcome. Infinite means infinite."

The judge acknowledges to the defense team that infinite does indeed mean infinite.

Essentially, the multiverse conception takes aim at the teleological principle, which posits that design and purpose found in human actions are inherent also as final causes in nature. The teleological use of the Anthropic Principle, suggests that the cosmos is fine-tuned to produce intelligent life.

Hawking is trapped. He can't banish the possibility of teleology and still argue that infinite possibilities exist. The Anthropic Principle does indicate teleological laws at work in our observable universe. The data is there so cosmologists have to deal with this issue if infinite possibilities exist.

Basically, the multiverse idea is an untested assumption so it lacks scientific merit. It simply adds unlimited roulette wheels so my number will come up. When speculative ideas are needed to make theories palatable, we have the right to question the credibility of such theories.

Roger Penrose has raised significant criticisms of the many worlds hypothesis to account for fine-tuning. Yet Hawking doesn't respond to any of these published critiques. To completely ignore the scrutiny of a physicist of Penrose's stature could be seen as deliberate blindness.

Invoking imaginary worlds and unsupported philosophical claims without doing the hard work of backing up the talk with valid data is hardly science. It reminds me of the unsupported realms of heaven and hell. In this regard, there's no difference between religion and science to an astute observer.

So far, *The Grand Design* has presented us with speculative ideas. Does M-theory with its unlimited universes represent the real world?

Taken at face value, Stephen Hawking's hypothesis seems to be half-baked. Any lawyer could reasonably argue that without evidence it's too far-fetched and deserves to be exposed as unverifiable.

7
Choose Your Universe

The title of Hawking's next chapter is "Choosing our Universe." (Yes, Virginia, that is the title). Obviously, the universe was already chosen for us before any human was born. So where is Stephen Hawking going with this?

We already know that Professor Hawking doesn't attach any importance to whether a model represents reality. In his book he stated, "According to model-dependent realism it is pointless to ask whether a model is real, only whether it agrees with observation."

The prosecutor now shifts his attention from the frailties of Stephen Hawking's grand design hypothesis to investigate the validity of his M-theory model which hinges entirely on whether it agrees with observation.

We can only observe a small part of our three-dimensional universe so is it possible to observe these limitless universes, all with different laws, and positing ten dimensions?

SIXTH CHAPTER

Dr. Hawking's Sixth Chapter portrays a few creation myths. He tells the creation story of the Boshongo people of central Africa and their God Bumba. Next, he describes the Mayan origins story.

It's easy to use the creation myths of unfamiliar African tribes and extinct cultures to contrast the modern creation story and enhance its appeal.

Of course, he doesn't use the Genesis story as an example – too many people might be offended. Nor does he use the Koran as an example – way too dangerous. But it's clear that he views the Bible and Koran creation stories as much a superstitious myth as any African origins story.

We may ask: How much is lost in translation from one language to another or from one culture to another? The Boshongo tribal chief might think, *how do I explain it to this English speaking person?* And how does the Boshongo language translate into English?

How does a physicist understand the story compared to an anthropologist? When the manuscript is complete it goes to an editor for editing and a publisher for marketing. Is the retelling of the Boshongo creation story to modern Americans and Europeans an accurate portrayal? Is it possible to understand the Boshongo story from a book the same way as people that live in that culture?

Even more difficult is retelling the Mayan creation story. The Mayan culture was wiped out in the 16th century by the Spanish Conquistadors. From where does Hawking get his information?

Perhaps from the diaries of Jesuit priests who accompanied the Conquistadors. But they had a preconceived agenda; convert the savages to Christianity. They claimed Mayan rituals and writings were the work of the devil in order to justify their ethnic cleansing.

It may be that Hawking salvaged some information from the four surviving Mayan codices, or from artifacts discovered at Mayan ruins by archeologists. Or, did he just Google online? Yet before any of this gets to Hawking it has already been interpreted and translated by others. Then he puts his own spin on it because he's an outsider. For him the story is absurd.

Ironically, Hawking's grand design creation story would sound equally absurd to the Boshongo people, the Mayan people, or any culture outside of the scientific tradition. Orthodox Jews and Muslims, for example, might say the modern creation story is ridiculous.

A creation story is appreciated by people within a culture because it relates to their immediate life. Persons outside the culture can never relate to it. For them, it's laughable.

Hawking utilizes the Boshonga and Mayan origins stories to make his point.

"Creation myths like these," he explains, "attempt to answer the question we address in this book: Why is there a universe, and why is the universe the way it is? Our ability to address such questions has grown steadily in the centuries since the ancient Greeks, most profoundly over the past century. Armed with the background of the previous chapters, we are now ready to offer a possible answer to these questions."

At last. Well, better late than never. Yet for some unknown reason his book goes off on a tangent. Instead of offering the possible answer he promised, Hawking distracts his readers with the following argument.

> "One thing that may have been apparent even in early times was that either the universe was a very recent creation or else human beings have existed only a small fraction of cosmic history. That's because the human race has been improving so rapidly in knowledge and technology that if people had been around for millions of years, the human race would be much further along in its mastery."

Is this tangent supposed to reveal why there's a universe and why it's the way it is?

Since Hawking brought it up let's look at other explanations. If a nuclear war takes place and a few thousand people have to start all over again, his logic would not apply.

A better hypothesis would acknowledge that ancient cultures were destroyed by natural catastrophes like tsunamis, earthquakes, and volcanoes. Or they might have destroyed themselves by their own "advancement." Civilizations like Rome disappeared due to war, depravity, and corruption.

According to Plato and his followers, Atlantis was an advanced civilization in antiquity that was wiped out, for some unknown reason.

Archeologists have discovered various lost cultures of antiquity underwater. Some examples include the Yonaguni Monument in the Japanese archipelago, the sunken cities below the Gulf of

Khambhat and off the coast of Dwarka in India, as well as the underwater city of Wanaku in Lake Titicaca, Peru.

These are forgotten chapters of the human story, proving our history is not one of uninterrupted progression. The fact that we don't remember simply means we are a species suffering from amnesia.

After the tangent, Hawking returns to the creation myth issue with another jab at people who accept the traditional story. "Bishop Usher [sic], primate of all Ireland from 1625-1656, placed the origin of the world at nine in the morning on October 27, 4004 BC."

Why in the world is this trivia in a serious physics book? And still Hawking is incorrect. Archbishop James Ussher dated the creation of the world as the night prior to Sunday October 23, 4004 BC. Many contemporary scholars offered calculations; even Isaac Newton proposed a date circa 4000 BC.

To be fair, Hawking does quote British physicist Sir William Thomson, also known as Lord Kelvin, who wrongly declared in 1884: "One thing we are sure of, and that is the reality and substantiality of the luminiferous ether."

What Kelvin held as knowledge is now considered ignorance. Thus, Dr. Hawking establishes that scientists can also be firm in their convictions which later turn out to be nonsense.

But why is there a digression towards false claims? Is this a portent of what's to come in *The Grand Design*?

In the 1960s the Steady State model of the universe became popular. Today, the public hasn't heard of this model. It's not even taught in schools. The fact is clear, as science marches on previous knowledge turns into ignorance.

At this point, Hawking wants to warn us not to make a mistake. He insists that it is:

> "...wrong to take the Big Bang literally, that is, to think of Einstein's theory as providing a true picture of the *origin* of the universe. That is because general relativity predicts there to be a point in time at which the temperature, density, and curvature of the universe are all infinite, a situation mathematicians call a singularity. To a physicist this means that Einstein's theory breaks down at that point and therefore cannot be used to predict how the universe began, only how it evolved afterward." [his emphasis]

This statement contradicts Hawking's PhD thesis which we briefly looked at earlier. He continues, "So although we can employ the equations of general relativity and our observations of the heavens to learn about the universe at a very young age, it is not correct to carry the Big Bang picture all the way back to the beginning."

Now it's clear. Einstein's theory can't predict the origin of the universe, and therefore Hawking's own PhD thesis is rendered null and void. Undaunted, he labors on.

> "Since we cannot describe creation employing Einstein's theory of general relativity, if we want to describe the origin of the universe, general relativity has to be replaced by a more complete theory. One would expect to need a more complete theory even if

general relativity did not break down, because general relativity does not take into account the small-scale structure of matter, which is governed by quantum theory."

Based on this remark, Hawking now turns to cosmic inflation to try and overcome the biggest hurdle in every origins story, a beginning point, or the beginning of time.

Here is the research on inflation. Theoretical physicist Alan Guth, then of M.I.T., first proposed the idea in the early 1980s suggesting that inflation was due to an anti-gravity force. During the first tiny fraction of a second after the Big Bang, he said, an anti-gravity field caused a runaway expansion of the universe. His hypothesis went beyond Einstein's relativity theory to invoke features of quantum theory.

Later in 1984, Guth and Paul Steinhardt wrote an article in *Scientific American* admitting the limitations of the idea, and that it afforded the opportunity for speculation:

> "The inflationary model of the universe provides a possible mechanism by which the observed universe could have evolved from an infinitesimal region. It is then tempting to go one step further and speculate that the entire universe evolved from literally nothing."[16]

Of course, if "the entire universe evolved from literally nothing" it would mean that "literally nothing" could have both mass and spatial extension.

Physics doesn't have an answer for what happened at the very beginning, how space was created out of literally nothing, if indeed that was the case. Presumably, a theory of quantum gravity could solve that problem.

By 2001, not much had changed in cosmology. *The New York Times* concluded in its science segment: "The only thing that all the experts agree on is that no idea works – yet."[17]

Thirty years after Alan Guth, physicists are still not sure how inflation happened. So in three decades little progress has been made. But for all the problems that inflation solved, it created others because it wiped out the possibility of knowing anything that existed previously. So now we have no idea what happened before inflation; we can only speculate.

Another problem: we are told that the inflationary expansion happened so fast that it far exceeded the speed of light, the maximum speed limit of the cosmos. Yet Hawking writes that the speed of light doesn't apply to the expansion of space itself.

This is interesting conjecture, but where is the data? Without proof it's only guesswork; what Richard Dawkins calls "blind faith without evidence." We revisit this issue in Chapter Nine of my book, *Cosmology on Trial*.

Physics still doesn't have a quantum theory of gravity, just ideas. Yet Hawking will justify his fusion of quantum theory and relativity theory via Alan Guth. Therefore, he writes:

> "...just as we combined quantum theory with general relativity – at least provisionally – to derive the theory of inflation, if we want to go back even further and understand the origin of the universe, we must combine what we know about general relativity with quantum theory."[18]

Did quantum events take place at the onset of the Big Bang? We can take the opinion of other cosmologists into consideration. Professor Paul Davies assures us that "quantum cosmology" – uniting the study of the universe with the study of atomic and subatomic systems – is an ambitious but questionable endeavor.

> "For a start, when dealing with quantum gravity, quantum mechanics has to be applied to space-time, not to matter, raising deep technical and conceptual problems."[19]

Again, Hawking doesn't seem to be bothered by the opinion of other physicists. Instead of considering the objections of colleagues, he prefers to go poetic, borrowing the metaphor of a flat world to press his point about the beginning of time using the inflation model.

> "The issue of the beginning of time is a bit like the issue of the edge of the world. When people thought the world was flat, one might have wondered whether the sea poured over its edge."

This quaint idea is an old wives' tale. Pythagoras wrote that Earth was a sphere as early as the sixth century B.C. Later, Aristotle and Euclid also wrote about a spherical Earth. At the height of the Roman Empire, Claudius Ptolemy wrote *Geography* stating that our planet was a sphere.

Even in the early 1200s a book titled *The Sphere* was published about a spherical Earth. By the 1300s this book was required reading in many universities of medieval Europe. It was still in use until the 1700s.

In 1991, University of California professor Jeffrey Burton Russell wrote *Inventing the Flat Earth*. His book clarified how the flat earth idea came to be accepted during the 1800s due to the inaccurate writing of Washington Irving, Jean-Antoine Letronne, and others.

Moreover, people who lived by the sea sustained themselves by seafood. They saw the boats go out daily and watched them disappear over the horizon. But they always returned with the day's catch.

On the boat, the fishermen would see their village disappear but everything was there as usual at evening. The same with conquering armies that went to sea, and persons who watched vessels sail in regularly with goods from other countries. Nobody thought the sea poured over the edge of the earth because no boat ever approached the edge of the earth.

If a frightened young girl saw father's boat disappear over the horizon, mother assured her that there's nothing to fear; the boat would return at evening.

So the idea of falling off the edge of the world is an invalid argument whose conclusion does not align with evidence, i.e. a *non sequitur*. If *The Grand Design* is meant to be a serious book on cosmology, why does Hawking refer to an old wives' tale with zero merit?

Yet he will use this false concept to present his theory about the beginning of time. He explains that time either has a beginning or it goes on forever. Einstein's general theory of relativity doesn't resolve this problem. If time had a beginning how did it start?

The modern origins story says time began with an exploding singularity. But that idea always had awkward problems, so Hawking will now present his new version of events where time behaves like space.

> "Although Einstein's general theory of relativity unified time and space as space-time and involved a certain mixing of space and time, time was still different from space, and either had a beginning and an end or else went on forever. However, once we add the effects of quantum theory to the theory of relativity, in extreme cases warpage can occur to such a great extent that time behaves like another dimension of space."

Factually speaking, Hawking's "extreme" case never happens. However, he proposes that it might have happened at the birth of the cosmos. His premise is that when the universe was at Planck size, a billion-trillion-trillionth of a centimeter, specifically a singularity, this is the scale where quantum theory can be used to explain events.

> "So though we don't yet have a complete quantum theory of gravity, we do know that the origin of the universe was a quantum event."

He posits that *if* the origin of the cosmos is governed by quantum theory and relativity theory, the result might be that time behaves like space; but only in extreme cases. Thus he speculates:

> "There were effectively four dimensions of space and none of time. There is no time at the origin of the

universe, because it's just another dimension of space."

This conclusion conveniently avoids the sticky question of how time began. It changed its nature into space. But this interpretation is reached by fuzzy logic at best. Hawking assumes that the quantum physics of sub-atomic particles with infinitesimal mass, will also work with infinite mass crunched into infinitesimal space – a singularity.

But is there data to support this assumption? Without data, or even a workable quantum theory of gravity, it's merely conjecture, not science.

Even if it could happen, it could only happen once. So this is not a law, nor is it a regular occurrence in nature. Do physicists accept this idea as empirical fact? Most people would consider it a miraculous event.

As hard as it is to believe, Stephen Hawking is prepared to stake his reputation on a miracle event that might have happened once – so it's not a law – if indeed it ever happened at all.

This is a bizarre idea, with no data, and a miracle to boot. But science does not support miracles. Ergo, the argument is dead; a still birth argument. In spite of this, Hawking will now accept his idea as given.

We have the right to ask, if time is postulated to behave like space in the beginning how did it again become like time? No answer is forthcoming. Yet Hawking assures us:

"We must accept that our usual ideas of space and time do not apply to the very early universe. That is beyond our experience, but not beyond our imagination, or our mathematics."

In Chapter Nine, I will discuss how mathematics can be manipulated to produce practically whatever result one desires. Hawking's statement is mostly imagination, as he himself implies.

What factually transpired at the onset of the creation when time behaved like space? Here is his answer:

> "The realization that time can behave like another direction of space means one can get rid of the problem of time having a beginning, in a similar way in which we got rid of the edge of the world."

This method of avoiding a sticky question by using an inappropriate argument is clearly fallacious in its reasoning. Let's follow his argument step-by-step:

1. the world having an edge creates a problem
2. but the world doesn't have an edge
3. therefore, the problem disappears
4. time having a beginning creates a problem
5. yet time is like a direction of space at the birth of the cosmos
6. therefore, the problem of time disappears

When people abandon logic and common sense they can blindly accept any theory. If we do not abandon logic and common sense we can easily see the error in an argument.

First, there never was a problem with the edge of the world. Nobody fell off the edge of the world and daily experience was the proof. There was no problem to get rid of except, perhaps, in the minds of blind followers.

Second, there is no problem with time. The dilemma is finding a proper explanation how time began *if* the Big Bang model represents reality. I deal with this extensively in my book *Cosmology on Trial*.

The third problem with Hawking's logic is that time is a completely different concept than space. Einstein fused them together mathematically as space-time. Thus, space-time is simply a mathematical space that is specified by space and time coordinates. To say time behaved *like* space is another kettle of fish entirely. That requires a huge leap of faith; a leap few scientists will stake their reputation upon.

Fourth, at the explosion of the singularity inflation ensues. The universe expands enormously within a split second, well beyond the influence of quantum physics which acts only on atomic scale. So in the initial split second after the Big Bang all quantum forces perish as inflation overrides all laws.

More to the point, it is highly questionable whether quantum forces could even act on the singularity due to its infinite mass. Quantum physics applies to infinitesimal mass.

Earlier in his book, Hawking stated that nothing can be said about a beginning using Einstein's equations. Consequently, relativity theory does not apply at the explosion of the Big Bang. So what are we left with?

Cosmology posits a singularity in which infinite density, infinite mass, infinite temperature and curvature, were all squeezed into an infinitesimal size, a Planck size. [that is 10^{-20}] An extraordinary situation. The singularity could not contain its cargo, and so it exploded. As soon as the singularity exploded the clock began ticking, according to some.

We may propose that the clock was already ticking prior to the explosion because, clearly, the singularity existed before it burst forth onto the scene. Yet no mention is made of the time factor relating to the existence of that singularity which contained all the matter, energy, and laws of the soon-to-be-born universe. It's similar to a gestation period.

I don't count the nine months in the womb, because the clock starts as soon as I'm born. However, for my mother, time begins nine months earlier. Scientists may not like this analogy because 1) it tends to lend credence to the traditional creation story, and 2) there is no science that can verify the singularity as a factual object.

However, in order for events to occur, time must already exist prior to, and independent of, an event. Or, time begins the moment an event begins. In either case, it follows that time already existed before the Big Bang because of the assumption of the existence of a singularity, which subsequently became our universe.

In his book, Professor Hawking fails to explain the mechanism that merges quantum theory and relativity theory which allows them to work in tandem. He skirts the issue and tries to assure us that, "...inflation has to be there in order for the theory to work, even though the explanation defies all other concepts of physics."

This admission is quite revealing. It has to be that way for the theory to work? Not because the real cosmos is that way? It's like Einstein adding his cosmological constant to fit a prevailing worldview without the prerequisite data.

Here again is proof that physicists shoehorn their ideas, hopes, and aspirations, to derive a theory even when the explanation "defies all other concepts of physics." By this definition, the theory itself is a miracle.

To make matters worse, Hawking will now offer the reader another imaginative example to buttress the "beginning of time" hypothesis. We will be asked to suppose that the beginning of the universe was like the South Pole of the Earth, with degrees of latitude assuming the role of magnitude.

"As one moves north," Hawking explains, "the circles of constant latitude, representing the size of the universe, would expand. The universe would start as a point at the South Pole, but the South Pole is much like any other point. To ask what happened before the beginning of the universe would become a meaningless question, because there is nothing south of the South Pole."

There is nothing south of the South Pole. This somehow proves we can dispose of time = zero?

Hawking's argument may sound profound sitting at his desk. He may draw a diagram of the Earth and the South Pole appears to be at the bottom and the North Pole at the top. But, if we go to the South Pole and the North Pole we will get a real world experience.

At the North Pole we may think we're at the top of the Earth and we can only move south in every direction. But at the South Pole, the situation is similar. You're not upside down. You might say we can only move north, but factually we can move in whatever direction we want.

"South" is merely an arbitrary direction, a semantic construct invented by man. By direct observation we can move in every direction at the South Pole exactly like at the North Pole, whether we name it south, north, or butterscotch. It makes no difference how we label it.

But let's continue with Hawking's analogy. The degrees of latitude eventually become smaller and end at the North Pole. The universe and time come to an end because there is nothing north of the North Pole?

For the empirical study of the real cosmos, this analogy has no connection to factual reality. It's certainly not astrophysics. "South of the South Pole" is just a semantic riddle that might have some sense as a brain-teaser. It has nothing to do with the physical reality that physicists study, or even a real life experience at the South Pole.

Hawking needs a much better analogy because "south of the South Pole" defeats his own case in terms of reason and logic. It's a matter of semantics, not of physics, because his South Pole argument doesn't hold in terms of physical reality. Nor does its application to the origin of the universe hold in terms of physical reality.

Both his analogies – the edge of the Earth, as well as its latitudes – are false. The first equates time and the edge of the earth on the basis that they are both problems, though other than being a problem for Hawking they have nothing else in common. The second likens latitudes with time, though they also have nothing in common, and since latitudes come to an end at the North Pole, time must too if the analogy holds.

His severe weakness in logic and philosophy leads him to erroneous and absurd conclusions that are just words, really, with zero merit for physical reality. And that's the failing of his book so far. However, Hawking truly believes his argument has merit:

> "The realization that time behaves like space presents a new alternative. It removes the age-old objection to the universe having a beginning, but also means that the beginning of the universe was governed by the laws of science and doesn't need to be set in motion by some god."

Factually, there is no data to support his "realization that time behaves like space." It's only an arbitrary decision to describe it that way. His description implied a miraculous event that defied all known laws of physics (if it even happened at all). So how was it "governed by the laws of science?" We need hard data to substantiate that time behaved like space.

Can science take credit for laws that existed billions of years before science existed? The laws of nature are universal laws, of which science has only discovered a small percentage. Remember, 96 percent of the energy and matter of the cosmos is unknown. Whatever laws govern this major portion of the cosmos are still unknown in the 21st century.

Once again, Hawking skirts the issue of the origin of the laws. Could it be south of the South Pole?

By sweeping it under the rug maybe he thinks it won't come up for discussion. This is like a hired housekeeper who accepts money for cleaning but sweeps dirt under the rug thinking no one will notice. It means the house of physics is not clean, at least not Hawking's house of physics where dirt is hidden.

It also indicates he's not being straightforward. This is the impression we are left with. Otherwise, he has a lot of explaining to do and a lot of unanswered questions to deal with. Where did the laws come from? How were they present before the creation of the universe? How were they formulated to govern the cosmos?

Science is still in a fledgling state. But comparable to the "facts" of Bishop Ussher and Lord Kelvin, Hawking's thesis may well be a source of jokes in the 22nd century. He denigrated philosophical reasoning at the beginning of his book, but from what we have analyzed it's now clear his ideas lack logic and scientific rigor to support his case.

If history is correct, and it usually is, we can safely conclude that within the next 50 years science will make many new discoveries. We'll have more knowledge, move forward, and leave the ignorance of today behind. But until then, today's ignorance will still be accepted as knowledge.

SEVENTH CHAPTER

Thus far in this courtroom drama, the prosecuting attorney has dealt Hawking a serious blow. *The Grand Design* book appears to be an incomplete study of an outdated perspective of 20th century physics, filled with wild speculations and false analogies.

Hawking's next chapter deals with the Goldilocks Enigma and is called "The Apparent Miracle" wherein he examines the bio-friendly fine-tuning of the universe.

I have briefly touched on the fine-tuning issue earlier, and I discuss it in detail in Chapter Three of *Cosmology on Trial*. Every physicist agrees that fine-tuning exists, so there's no need for further discussion here. But we may consider some rhetorical questions raised by Professor Hawking at the end of his sixth chapter.

"What are we to make of this fine-tuning?" he asks. "Is it evidence that the universe, after all, was designed by a benevolent creator? Or does science offer another explanation?"

Well, we've been waiting for another explanation since page one. Hopefully, we'll find that explanation in the final chapter of his book. Thus far, he has not credibly resolved the issue of a beginning for the modern creation story. Will we finally discover his vision for a unifying theory?

8
The Game of Life

The last chapter of Hawking's book is titled "The Grand Design." At last we come to his grand vision of a unifying theory, hopefully.

EIGHTH CHAPTER

He begins with the following statement: "regularities in the motion of astronomical bodies such as the sun, the moon, and the planets suggested that they were governed by fixed laws..."

Again, the issue of the origin of the laws is sidestepped, but moving on, now he declares that there "must be a complete set of laws that, given the state of the universe at a specific time, would specify how the universe would develop from that time forward. These laws should hold everywhere and at all times; otherwise they wouldn't be laws."

Of course we anticipate that he will present a multiverse conception, with endless universes and infinite sets of laws. Even in our observable universe there are different sets of known laws and not all of them "hold everywhere and at all times." We have quantum physics which only works for sub-atomic particles, and Newton's laws which do not apply to sub-atomic reality.

Undeterred by any sense of contradiction Hawking continues:

> "At the time that scientific determinism was first proposed, Newton's laws of motion and gravity were the only laws known. We have described how these laws were extended by Einstein in his general theory of relativity, and how other laws were discovered to govern other aspects of the universe."

It looks like we will now get to the essence of Hawking's thesis.

"The laws of nature tell us *how* the universe behaves, but they don't answer the *why* questions that we posed at the start of the book.

> Why is there something rather than nothing?
> Why do we exist?
> Why this particular set of laws and not some other?"

My own *why* question is why we had to wade through all the other stuff when we could have gone straight to this immediately? Regrettably, the following explanation provides no answer to the questions:

> "Some would claim the answer to these questions is that there is a God who chose to create the universe that way. It is reasonable to ask who or what created the universe, but if the answer is God, then the question has merely been deflected to that of who created God. In this view it is accepted that some entity exists that needs no creator, and that entity is called God. This is known as the first-cause argument

for the existence of God. We claim, however, that it is possible to answer these questions purely within the realm of science, and without invoking any divine beings."

Hawking has addressed the question of who created God with the first-cause argument. I'll briefly clarify the 'who created God' conundrum with a more logical explanation.

God is defined as a supernatural entity that exists in a supernatural realm that is eternal. There is no beginning or end to this domain. It's described as supernatural because the effects of time – which we perceive as birth, growth, aging, disease, and death – would be absent in such an environment. The question who created God makes no sense in an eternal realm with no beginning or end.

Does such a realm exist or is it pure fantasy? Well, according to Hawking's account of M-theory, there are infinite universes with infinite possibilities. With no data to prove supernatural realms do *not* exist, then they can't be excluded from the infinite possibilities.

However, our interest is in the explanation of the science creation story, so I'm looking forward to that. But, there is yet another digression by Hawking. Instead of answering his own questions quoted above, he refers to the concept of model-dependent realism that he introduced earlier.

"...our brains interpret the input from our sensory organs by making a model of the outside world. We form mental concepts of our home, trees, other people, the electricity that flows from wall sockets, atoms, molecules, and other universes. These mental

concepts are the only reality we can know. There is
no model-independent test of reality."

He claims we are dependent on a model to understand the
universe because "our brains interpret the input from our
sensory organs by making a model of the outside world." Does
that prove there is no test for reality independent of a model?

We can see the results of experiments proving the existence of
atoms and molecules which allow us a glimpse into their
nature. Therefore, the concurring opinion of experts is an
indication of the reality of a theory.

Similarly, we all see and utilize rooms with wall sockets. The
concurring opinion of people in a room who see a wall socket, is
a sure indication that the socket is real.

Conversely, if a person in a room sees a ghost of his mother and
others do not, the scientific conclusion would be that it was not
real, just a figment of his bereaved mind. Following this logic,
we are not required to accept one person's test of reality.

When *every* person sees the same thing and describes it the same
way, then we may conclude that it is what it is, beyond our own
mental construct, or model. This test of reality is independent of
a model.

Yet Hawking makes another false claim: "It follows that a well-
constructed model creates a reality of its own." His reasoning
goes like this: Reality is dependent on a well-constructed model.
I have such a model. Therefore I have reality.

We have shown that reality can exist independently of an
observer. If I don't see it someone else will. In fact, a well-
constructed model could just as readily create an *illusion* of its

own. The board game "Monopoly" is an illusory model of genuine real estate. How to decipher the real from the illusory? That's the real issue.

In this section of his book, Hawking has resorted to philosophical concepts but not scientific methodology. We deserve answers from the realm of physics. He has kept us waiting so long that I'm having doubts he can deliver.

Why is it difficult for Hawking to give straightforward answers? If anyone else thinks Professor Hawking is going round in circles, then I'm not alone.

THE GAME OF LIFE

All of a sudden, the Grand Design becomes a Grand Disappointment. Dr. Hawking is going to model his theory of creation on a board game invented in 1970 by John Conway. It's called *The Game of Life*.

I won't go into the boring details, but Hawking's model of the universe will mirror the conclusions derived from the rules of this game created by Conway. It sounds unbelievable, yet Hawking admits this in his book so it's not my misconception. Is it because "a well-constructed model creates a reality of its own?"

He takes up 13 of the final 19 pages of his book to explain Conway's logic and rules of the game. In these 13 pages he uses the *Game of Life* universe to extract a model for our real-world physical universe. Using the attributes of Conway's model, he extrapolates them to the physical laws of nature.

"As in our universe, in the *Game of Life* your reality depends on the model you employ." As in our universe? This is the fallacy

105

of Begging the Question – the circular argument wherein the conclusion is already in the premise. What is meant to be proven has already been accepted.

Although Hawking accepts his premise as a given, we have shown that a well-constructed model like "Monopoly" could just as easily create an illusion of its own, and often does for the serious player.

> "The example of Conway's Game of Life shows that even a very simple set of laws can produce complex features similar to those of intelligent life. There must be many sets of laws with this property. What picks out the fundamental laws (as opposed to the apparent laws) that govern our universe? As in Conway's universe, the laws of our universe determine the evolution of the system, given the state at any one time. In Conway's world we are the creators – we choose the initial state of the universe by specifying objects and their positions at the start of the game."

A simple point of contention: Hawking can't choose the initial state of the universe (as he can in Conway's game) because the initial state of the cosmos, as well as the laws that govern it, was already "chosen" at the Big Bang. The original conditions and the basic laws of physics already limit what physical systems can and cannot do.

It makes no difference whether the cosmos began with a Big Bang, a Little Bang, or Many Bangs. Nobody can adjust the rules of the universe because the laws are pre-determined.

Our interest is solving the puzzle of how everything came into being. Professor Hawking promised us this was forthcoming, but now he's extrapolating from Conway's board game.

Why doesn't Hawking come up with original ideas? Always borrowing from others reduces his stature and reputation. Perhaps John Conway should get credit for the Grand Design theory.

Finally, in the last three pages of his book Stephen Hawking presents his explanation.

> "In a physical universe, the counterparts of objects such as gliders in the Game of Life are isolated bodies of matter. Any set of laws that describes a continuous world such as our own will have a concept of energy, which is a conserved quantity, meaning it doesn't change in time. The energy of empty space will be a constant, independent of both time and position. One can subtract out this constant vacuum energy by measuring the energy of any volume of space relative to that of the same volume of empty space, so we may as well call the constant zero."

He has arbitrarily assumed that the energy of empty space is a zero constant, yet the presence of dark energy and dark matter is necessary to explain the expanding universe, even in empty space. What follows is clearly deceptive:

> "If the total energy of the universe must always remain zero, and it costs energy to create a body, how can a whole universe be created from nothing?"

107

"If" is a small word with a huge meaning. From assuming that the energy of any volume of space is a zero constant relative to that of the same volume of empty space, he concludes that "the total energy of the universe must always remain zero."

My simple point is this: since 96% of the universal matter and energy is unknown no one can assume it's actually zero without presenting data as evidence. Physicist J. J. Thomson, who discovered the electron, was very succinct about the role of using mathematics in physics:

> "We have Einstein's space, de Sitter's space, expanding universes, contracting universes, vibrating universes, mysterious universes. In fact the pure mathematician may create universes just by writing down an equation, and indeed if he is an individualist he can have a universe of his own."[20]

Stephen Hawking is indeed an individualist. By assuming first that the energy of empty space is a zero constant, he concludes that the total energy of the universe must also be zero by positive and negative energies canceling each other.

Furthermore, in an attempt to make them sum to zero, he arbitrarily assigns them positive and negative values. Thus, by playing with the numbers he comes to this conclusion:

> "...there must be a law like gravity. Because gravity is attractive, gravitational energy is negative. One has to work to separate a gravitationally bound system, such as the earth and moon. This negative energy can balance the positive energy needed to create matter."

His assumption that the positive energy required to produce matter generates an equal and opposite negative energy called gravity, is based on this: "One has to work to separate a gravitationally bound system, such as the earth and moon."

Again, this is deceptive. What keeps the moon and earth separate, though gravitationally bound, is the tangential velocity of the moon, which is the right speed for it to remain in constant orbit. A faster or slower speed would cause the Moon to fly off into space or crash into the Earth. This has nothing to do with the "positive energy needed to create matter."

Moreover, in Hawking's model gravity *already* exists and it can create matter. It's like the game of "Monopoly" where you already start out rich. No inkling from him how gravity came to exist, nor what sort of force grounds the laws so that they hold in all cases and continue to govern the universe.

All questions about universal laws are left unresolved, yet he continues:

> "A body such as a star will have more negative gravitational energy, and the smaller it is (the closer the different parts of it are to each other), the greater this negative gravitational energy will be. But before it can become greater than the positive energy of the matter, the star will collapse to a black hole, and black holes have positive energy. That's why empty space is stable. Bodies such as stars or black holes cannot just appear out of nothing. But a whole universe can."

Dr. Paul Davies explains why "empty space" can't be empty. The answer is "provided by quantum mechanics, which predicts that even apparently empty space is teeming with virtual particles."[21]

So the nothingness, or the zero constant that Hawking bases his model on, doesn't seem to hold water. He claims that matter and black holes have positive energy so they can't appear out of nothing. But the universe they are within, can. Why?

"Because gravity shapes space and time, it allows space-time to be locally stable and globally unstable."

Hawking claimed earlier at the beginning of the universe, "there were effectively four dimensions of space and none of time." If there was no time at the very beginning (because it behaved like space) there couldn't have been space-time as both space *and* time coordinates must be specified. The above explanation is simply another contradiction.

"On the scale of the entire universe, the positive energy of the matter *can* be balanced by the negative gravitational energy, and so there is no restriction on the creation of whole universes." [his emphasis]

He claims it's possible that the energy can cancel out to zero, although no data is given as supporting evidence. But even if that is the case how does this balance remove restrictions on the creation of whole universes? And how does he come to the following unproven conclusion?

"Because there is a law like gravity, the universe can and will create itself from nothing..."

If positive and negative energies along with gravity were *a priori* causing agents, then clearly the universe did not create itself from nothing. That is not a coherent conclusion. To say that something creates itself from nothing means that it's self-created with no help from other things that already exist.

110

Moreover, from his premise that positive and negative energies cancel each other out it does not follow that the universe "can and will create itself from nothing" with every law perfectly fine-tuned for life thrown into the bargain.

The conclusion of Hawking's incoherent doctrine lies in the following sentence:

> "Spontaneous creation is the reason there is something rather than nothing, why the universe exists, why we exist."

The term "spontaneous creation" means that "nothing" generated "everything" which continues the irrational idea of self creation.

Is there any data to support spontaneous creation (positive and negative energies arising from nothing) as opposed to god creation? Otherwise Stephen Hawking consigns the scientific creation story to the same category as the traditional story – blind faith with no evidence.

From the perspective of logic and reason, saying the universe creates itself from nothing sounds more far-fetched than having a god as the creator. At least the god conception denotes an intelligent creating agency.

When there are two alternatives, is there a mechanism, or rationale, for choosing one over the other beyond an arbitrary or whimsical decision? Here is Einstein's conclusion:

> "When two theories are available and both are compatible with the given arsenal of facts, then there are no other criteria to prefer one over the other except the intuition of the researcher."[22]

Professor Hawking's book presents no "given arsenal of facts." There are no criteria to support his creation myth over the traditional creation myth. His preference is just an arbitrary choice.

Anticipating a fire storm of outrage, Hawking presents this explanation:

> "Why are the fundamental laws as we have described them? The ultimate theory must be consistent and must predict finite results for quantities that we can measure. We've seen that there must be a law like gravity, and we saw in chapter five that for a theory of gravity to predict finite quantities, the theory must have what is called super-symmetry between the forces of nature and the matter on which they act. M-theory is the most general super-symmetric theory of gravity. For these reasons M-theory is the only candidate for a complete theory of the universe. If it is finite – and this has yet to be proved – it will be a model of a universe that creates itself. We must be part of this universe, because there is no other consistent model."

No comment from me. If what's been presented "has yet to be proved" this is condemnation enough. Famed scientist Robert Matthews sums it all up:

> "Take quantum theory, the laws of the subatomic world. Over the past century it has passed every single test with flying colors, with some predictions vindicated to ten places of decimals. Not surprisingly, physicists claim quantum theory as one

of their greatest triumphs. But behind their boasts lies a guilty secret: they haven't the slightest idea why the laws work, or where they come from. All their vaunted equations are just mathematical lash-ups, made out of bits and pieces from other parts of physics whose main justification is that they seem to work."[23]

The Grand Design certainly let us down. Resorting to a board game invented in 1970 doesn't sound like 21st century physics to me.

Stephen Hawking's explanation that positive and negative energies balanced each other and along with gravity supplied some stimulus for spontaneous creation, contradicts the entire basis and definition of nothingness.

Besides, why would a singularity in a state of rest suddenly release its contents, thereby creating the universe? Some impetus at time t=0 must have been the force to provoke the change according to known laws of physics.

The traditional story proposes an impetus from God which begins the creation. This idea remains unproven, but it does provide an active force to answer the dilemma. Cosmology's failure to provide its own rationale puts it at a distinct disadvantage.

Hawking's explanation of a "grand design theory" cries out for a response about consciousness, which is ubiquitous on planet Earth, but he doesn't acknowledge it. He ends his book with this comment:

"M-theory is the unified theory Einstein was hoping to find. The fact that we human beings – who are ourselves mere collections of fundamental particles of nature – have been able to come this close to an understanding of the laws governing us and our universe is a great triumph. But perhaps the true miracle is that abstract considerations of logic lead to a unique theory that predicts and describes a vast universe full of the amazing variety that we see. If the theory is confirmed by observation, it will be the successful conclusion of a search going back more than 3,000 years. We will have found the grand design."

Did you notice the disclaimer? "If the theory is confirmed by observation" informs us that there is no evidence, no data.

Remember what Hawking stated in his third chapter. "According to model-dependent realism it is pointless to ask whether a model is real, only whether it agrees with observation."

If the reader assumed that the Grand Design model *would* agree with observation, it's another disappointment.

Factually, there is zero observational data to support his "spontaneous creation" model. Therefore, by his own statement, its validity is flawed. Consequently, the grand design theory is mostly conjecture and barely even science.

Our mock prosecutor comments, "It's now clear why Roger Penrose said Hawking's M-theory was hardly science – *It's a collection of ideas, hopes, aspirations; it's not even a theory.* Case closed; game over."

9
Faulty Theories& Fuzzy Math

If you're following the trial, at this point you are probably dismayed by what we have discovered. We always believed that science dealt with facts which we could trust. Yet, a prominent cosmologist like Stephen Hawking presents information that is as speculative as any tribal creation myth.

I can now understand that Hawking's brilliance is not in his power of logic or reason. Rather, it's that he has been able to bring physics to the masses. His style is concise and understandable. He adds a dash of humor which I think many appreciate, though some might consider corny.

We also have to acknowledge his determination to stay alive and continue his work after he was written off by doctors that he would never live to finish his PhD. But it's quite disconcerting that an investigative journalist can dismantle a theory of cosmology by a person acclaimed to be brilliant.

Of course, all credit must go to the scientists offering their testimony. They work hard to resolve contradictions and unravel problems. Brave physicists put their career at risk by speaking up when they find data that contradicts the *status quo*.

CRITICAL ANALYSIS

Professor Hawking's book, *The Grand Design*, seems to be his swan song – an attempt to be recognized for providing a unique contribution. However, the book spoils his aspiration to be recognized by history as a great contributor to the field of cosmology.

Instead of giving us solid physics, his book simply served up conjecture and speculation. No grand design of the universe shone forth. No brilliant physics except for a rehash of what Feynman, Guth, Einstein, and even Conway presented years ago. There's nothing new for the cosmology story that hasn't been discussed before, in greater detail, by other physicists.

We may question whether Hawking's conclusions are even original. For example, in the March 1984 edition of *New Scientist* magazine an article by physicist Edward P. Tryon was titled, *What Made the World?*

On page 15 he wrote: "I proposed that our Universe had been created spontaneously from nothing, as a result of the established principles of physics." Tryon says he originally proposed this in 1973.

We already quoted Alan Guth and Paul Steinhardt who later echoed the same idea as Tryon, and now Hawking. So his conception is not original.

Historically, the idea of the self-creation of matter espoused by Hawking's grand design theory has been part of an underlying direction of modern cosmology since the time of Charles Darwin.

Let's suppose the universe could spontaneously erupt from literally nothing. Then "literally nothing" can extend itself into space, mass, energy, gravity, consciousness, love, and infinite varieties of life. Such reasoning suggests that ideas in physics and cosmology are deteriorating into utter nonsense.

Alternatively, we can construe a more reasonable explanation by way of a higher dimensional reality. The so-called "nothing" could not be comprehended simply due to a lack of knowledge, which led us to an incorrect conclusion: that which can't be detected, or figured out, is taken to be nothing.

So far we have seen that most of Hawking's book is philosophically, or theoretically, based. What remains is conjecture. There is no observation, no experimentation, and no objective testing to verify the proposed Grand Design theory. Because he has no empirical data he resorts to philosophy throughout his book, but he does so by appealing to the use of misleading and inaccurate metaphors.

By saying that the universe can spontaneously arise from nothing Hawking misuses the meaning of "nothing" which is the absence of anything at all. Again, nothing means that "no thing" exists to have any properties that could generate any cause or effect. So nothingness could never be the source of anything, what to speak of the entire universe.

Nor is empty space consistent with nothingness, because space is no longer considered to be empty. The quantum vacuum is also not "nothing" – it is a sea of fluctuating energy.

118

Because he uses the word "nothing" in an inaccurate sense philosophically, Hawking comes to an erroneous conclusion. And it's hard for me to believe that he has not made this connection himself.

We have cause to question: did he deliberately choose to misuse the meaning of "nothing" to justify his hypothesis?

His book relies upon grandiose and baseless philosophical pronouncements devoid of scientific rigor. Why didn't Hawking give rigorous proofs and substantiate M-theory through logical hypotheses? He could have included research from Ed Witten. He should have included perspectives from other theoretical physicists in the field. But he didn't. So we took him to the court of public opinion.

Physicist Paul Davies is quite frank about the relevance of M-theory: "...it uses branches of mathematics which are not only extremely abstract, but also extremely obscure. In fact, some of the mathematics had to be invented along the way."[24]

The obscure and invented mathematics leaves M-theorists with no reality check. It reminds me of the ugly stepsister trying to fit her big foot into Cinderella's delicate glass slipper.

Professor Hawking concludes that no god is necessary for the universe to have come into being and to exist. Instead, gravity and the laws of nature are necessary for the universe to come into being and exist. But without resolving the enigma of the origin of gravity, the origin of the laws, and why they always work, he leaves a huge gap unexplained.

A theory can't have such a major flaw, a sizable gap left unresolved and unaccounted for, and still maintain its credibility.

Another problem is his statement about the positive and negative energy of the universe canceling out. "On the scale of the entire universe, the positive energy of the matter *can* be balanced by the negative gravitational energy, and so there is no restriction on the creation of whole universes." [his emphasis]

Does he mean that today's matter and gravity creates new universes? Wasn't his grand design theory intended to resolve the initial creation issue? Or, does the positive energy of matter and the negative energy of gravity exist before the cosmos is created? If so, then how are those nothing? This is a huge gaffe. Does Professor Hawking even review what he has written?

Remember, he is trying to solve the issue of the initial beginning of creation when there was nothing in existence – not time, not space, not matter, not gravity, and therefore no energies that can oppose each other.

The stumbling block of Hawking's self-creation of universes is that matter, energy, gravity, and M-theory suddenly appeared from nothing, and they were already present to allow for spontaneous creation. But how they came into being, and where did they come from? That vital question is never addressed.

Furthermore, if we accept the universe came into being from nothing that begs the question: why don't other things do the same? We should be able to see other things simply appear from nothing if this is part of naturalism.

There is no experience of anything ever arising from nothingness, yet Dr. Hawking claims this for an extreme case, the universe itself. But that claim can never be verified or falsified. It can only be accepted on blind faith.

The bio-friendly nature of the cosmos is another major problem for Hawking. He explains it away with unlimited roulette wheels; the many worlds idea. But it's not convincing.

The multiverse can never be verified or falsified because it lies outside of our observable universe. This idea is similar to the unseen realms of heaven and hell that religions describe which also lie outside our observable universe.

These flaws make his book look like an amateur attempt at scientific philosophy; or an abuse of philosophy. It's like a manual for the professional gambler: "bet on multiple wheels to improve your chances." Needless to say, such desperate grasping at straws falls far short of the renowned classical intellects enshrined in the halls of science.

To accept Hawking's contention that the universe created itself, we would first have to consider whether the universe is tapping into a universal intelligence or some type of cosmic consciousness. Why? Because mathematical laws, and laws of nature, indicate some sort of underlying intelligence.

The principle of entanglement, whereby particles communicate with each other across the universe, indicates connectivity with some kind of awareness over vast distances.

It's logical and reasonable to attribute advanced consciousness to laws which describe and define nature. But Hawking wants us to accept that everything arose spontaneously from nothing, from zero. He claims a "grand design" yet his model doesn't even consider the subject of consciousness.

Let's ponder another thought experiment. The nature of my body, from a quantum physics perspective, is almost entirely space. My skeletonis also mostly space. The nature of matter is that even solid rock is mostly space with tiny hubs of mass/energy we call atoms.

The universe is also mostly space with tiny hubs of mass/energy we call stars and planets. Its composition is qualitatively similar to the human body, although quantitatively quite different.

The human body is a conglomeration of atoms that follows the laws of nature. At the same time it follows the dictates of a conscious will to serve a specific purpose. The question whether there is a will in the universe, could be evidenced by a purpose coming about in the universe.

Since the universe is so bio-friendly one obvious purpose, universally, could be the creation of life. From the definition that my body is a conglomeration of atoms pervaded by consciousness, we get the idea that the universe is also a body of matter that might be pervaded by consciousness, perhaps even directed by will.

Because we can't see cosmic consciousness doesn't mean it's not there. Not being able to see dark matter and dark energy doesn't mean they don't exist. The fact that I have a body which serves my purpose by my unseen consciousness is proof that I exist. The idea of a cosmic consciousness might be how most people define God, which could lead to a synthesis of science and theology.

Consider this: If no link exists between life, consciousness, and the cosmos, why are we discovering laws that work in harmony to produce life plus everything required to support life? Why are scientists searching for a link? Or, shall we conclude that the origin of natural laws is random nuclear fusion?

If physicists accept the challenge of investigating the role of consciousness, many troubling issues might be better understood. This idea is exactly what the original quantum physicists realized during the 20th century.

From the perspective of Max Planck, the father of quantum physics, much of the 20th century was wasted because later quantum theorists failed to expand on his work. He demonstrated that consciousness affects the results of quantum physics experiments. In 1918, Planck received the Nobel Prize for his contribution to science.

After a lifetime dedicated to physics, Planck's work convinced him that consciousness was a primary and pre-existing force. He explained it this way:

> "As a man who has devoted his whole life to the most clear-headed science, to the study of matter, I can tell you as a result of my research about atoms this much: There is no matter as such. All matter originates and exists only by virtue of a force which brings the particles of an atom to vibration and holds this most minute solar system of the atom together. We must assume behind this force the existence of a conscious and intelligent mind. This mind is the matrix of all matter."

Is Planck's version on the right track? Consider one of the fundamental principles of classical physics – that a change can only come about by an applied force.

For example, the theory of evolution posits that things change and evolve. In order to shift from a settled state there must be some impetus to launch the change. According to the prevailing theory, survival is the motivating stimulus that pushes evolution. Adaptation provides the momentum for change that ensures survival. In this way the forward motion of evolution persists by a force – the conscious will to survive.

Likewise, without some sort of stimulus nothingness would remain as it is. How would the universe come into being from nothing unless an applied force acted upon it?

In a *Scientific American* editorial Philip and Phylis Morrison present a brief history of the hoard of cosmological theories that were later overturned.

> "We simply do not know our cosmic origins; intriguing alternatives abound, but none yet compel. We do not know the details of inflation, nor what came before, nor the nature of the dark, unseen material, nor the nature of the repulsive forces that dilute gravity. The book of the cosmos is still open. Note carefully: we no longer see a Big Bang as a direct solution. Inflation erases evidence of past space, time, and matter. The beginning – if any – is still unread. It is deceptive to maintain so long the very term that stood for a beginning out of nothing. The chanteuse will compose a clever new song once the case is clear."[25]

An observation by Mark Twain appears appropriate for the grand design theory: "There is something fascinating about science. One gets such wholesale returns of conjecture out of such a trifling investment of fact."[26]

Science philosopher Karl Popper summed it up emphatically by stating that, "A scientific theory neither explains nor describes the world; it is nothing but an instrument."[27]

Professor Hawking is a person striving to say something brilliant, but comes up short. He really wants us to accept that his M-theory explanation is "the *only* candidate for a complete theory of the universe." [his emphasis]

"Only" means everything else is rejected. There can be no other candidate. Yet his grand design theory is comprised only of conjecture and contradiction. It's unobservable, untestable, and *only* a grand delusion. [my emphasis]

Hawking has failed to see that his creation account is as indeterminate as any traditional creation story. In every creation story the fundamental question is, what started it all? This enigma is also at the root of Hawking's theory.

The traditional creation story claims that a supernatural God of unlimited intelligence and power started it all. There is no empirical evidence to support the assertion that a conscious being far beyond human intelligence and ingenuity could have brought it all together and set it in motion. Yet, taken at face value, the explanation of an initial force is reasonable enough.

We know that intelligence and ingenuity are required to create and set into motion anything humans are able to accomplish. Based on experience, it's a plausible step to attribute extreme fine-tuning for life to a highly advanced intelligence from another dimension of reality.

Of course, this is not an attempt to substantiate any particular god of any particular religion. It only points to advanced intelligence that could have existed prior to life as we know it. After all, the plausibility of consciousness being on a cosmic scale, based on conclusions from the double-slit experiment and the phenomenon of particle entanglement, has already been noted by physicists such as Wigner and Planck.

In a different sense, Richard Dawkins alludes to this in a filmed interview: "It could be that at some earlier time, somewhere in the universe, a civilization evolved by probably some kind of Darwinian means, to a very, very high level of technology, and designed a form of life that they seeded on to, perhaps, this planet. That is a possibility and an intriguing possibility."

Panspermia is the science that considers life on Earth was imported via meteorites, comets, or asteroids. Dawkins continues to support this idea.

"I suppose it is possible that you might find evidence for that. If you look at the D cells of chemistry or molecular biology, you might find a signature of some sort of designer, and that designer could well be a higher intelligence from elsewhere in the universe."

So an advanced consciousness life form is a viable possibility within our vast universe. Dawkins concludes: "But that higher intelligence itself would have had to have come about by some explicable, or ultimately explicable process. It couldn't have just jumped into existence spontaneously. That's the point."

That is indeed the point. For Stephen Hawking to present that the universe just jumped into existence spontaneously from nothing, is not an explicable or plausible proposal. His grand design theory is an example of faith with not a shred of data. We know this is the definition of Scientism.

We don't have to accept Hawking's version blindly and just surrender our native intelligence, sound judgment, logic and reason. A more sensible approach is to take the conclusions of other expert physicists and mathematicians into account.

After reading *The Grand Design*, one can't help thinking about Andersen's tale of the vain Emperor. The grand design theory turned out to be a grand deception.

Hawking and Mlodinow appeared less like scientists and more like swindlers trying to sell us the Emperor's new cosmos.

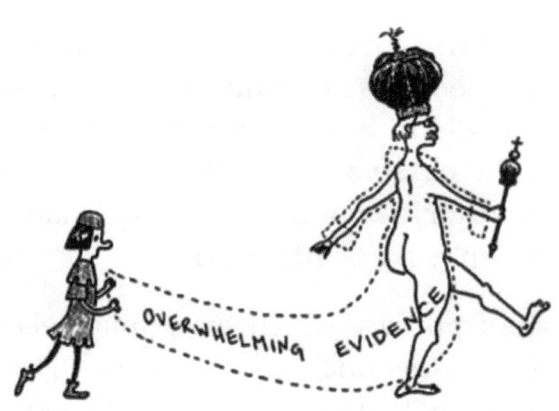

Credit: anonymous political cartoonist emperorswithoutclothes.com

Current speculations about a theory of everything have pushed science beyond the wildest imagination of science fiction. Is M-theory a mirage of mathematics, or can it lead us to an ultimate theory of everything?

If a mathematical formula accurately simulates a known law, then physicists can calculate on paper what *might* happen in real life. However, because they work in an office, not a laboratory, they're more like armchair pilots playing with flight simulators.

They predict various outcomes and look for evidence to support their ideas. To get the desired result they manipulate the numbers in diverse ways, producing various results, some acceptable and others nonsense. When the ideas are not confirmed, they're just dropped. But when they do find something to confirm an idea they say the theory predicted it.

Thus, using a successful theory, a physicist can predict an answer for a problem by simply applying the relevant mathematical equations that model the laws under examination. And that's why they say the theory predicted the outcome.

Mathematicians have already informed us that the use and function of mathematics beyond its traditional scope of explaining observations, has evolved into creating models that have no basis in observation of the real world. Thus, the scientific creation story is not as certain as it's taught. This calls the perspective of physicists into question.

Even the areas of physics, which the public believes are well understood by the scientific community, are gradually being revealed as clouded with doubts and disagreements.

Of course, the purpose of this book is only to inform the public of the anomalies in the modern creation story. It is not intended to be a complete treatise on the state of cosmology.

Please browse through the endnotes which comprise a bibliography directing inquisitive readers to the writings of the luminaries of astrophysics. For an in-depth study it's always best to read directly from the eminent physicist's original writings.

Another Book by the Author

Cosmology on Trial: Cracking the Cosmic Code

Amazon Bestselling author, Pierre St. Clair, examines present models of the universe like an attorney in a court of law, without complex jargon.

The author's manner is, "Just the facts please. Show me the evidence."

Cosmology On Trial presents the remarkable findings of his study and introduces readers to unexplained and unsolved mysteries of the universe.

Link: http://amzn.to/1UBj5ni

Acknowledgements

I began this investigation into Big Bang cosmology by a chance discovery of the book *The Goldilocks Enigma*. Therefore, I want to recognize the inspiration I received from the author, British physicist Dr. Paul Davies. My story begins and ends with insights from Professor Davies.

Of course, I must also give immense credit to my beautiful young wife, Chris, who encouraged me along the way, even when the going got tough and it looked like we were on a wild goose chase that would never end.

I sent the first draft of the book to my friend Joshua Wulf, who made valuable comments, including that he was amazed that I was a physics buff.

I sent the second draft to physicist Mauricio Garrido who was recommended by a friend. Mauricio guided me to papers and ideas beyond what I had found by my own research. He also clarified certain concepts that really needed amplification.

Although I had a manuscript that was logically and scientifically solid, I still needed an editor to make it understandable to the general public not well versed in cosmology. I was led to Noelene Musumeci, a philosophical editor who reviews all sorts of manuscripts. She told me she knew hardly anything about cosmology, but would give it a read. I responded that she would represent the uninformed public that my book hoped to inform, so her input would be invaluable. And indeed it was.

I now had a strong document that needed a cover, promotion, and sales. As luck would have it, I bumped into an old friend Alister Taylor who helped me with book publishing. We found a designer who did an expert layout for the cover. Next, we developed a marketing plan and the book became a reality. Now you know the whole story.

About the Author

Bestselling author Pierre St. Clair focuses on anomalies in cosmological theories.

Recently he won the prestigious Sir J. C. Bose Award for Investigative Journalism in the field of Science and Technology for his book Cosmology on Trial.

Although born in London, at five years of age his family emigrated to Canada. His mother was a cultured lady who sent him for violin and music theory lessons when he was only six. As a teenager his interest in classical violin dwindled. He bought his first guitar, taught himself how to play, formed a band, and began performing locally.

Within a few short years he was touring throughout Canada and the US. Music was his first love, and remains so.

His journalism career began partially as a result of his college girlfriend's mother. The daughter of a missionary in Shanghai, she had grown up in China and pursued an interest in the occult sciences, including palmistry. One Saturday afternoon she asked to see his hand.

After a few minutes study she pronounced that he had a talent for writing and predicted, "One day you will become an author."

Soon that writing talent manifested.

"My interest in writing, and especially research, increased. I loved digging behind the scenes to find out what was really happening beneath the headlines. That's how I became an investigative journalist.

"I've traveled to 76 countries in the course of investigating a variety of stories. Travel has been my best education about culture, tradition, and food. Yes, I also have a gourmet lifestyle dining on an assortment of traditional cuisines worldwide."

Please visit his Author Page:

http://www.amazon.com/author/pierrestclair

Connect With Me

If you're reading this, it means you purchased my book and have read it.

My heartfelt thanks for your support.

So what did you think? Let me know what you liked, or what you thought was lacking.

I will personally read and respond to questions and comments on my Facebook page.

Here is the URL: www.facebook.com/cosmology.on.trial

Please send me your feedback because it's an opportunity to make this a better book for future readers:
cosmologycrisis@gmail.com

I look forward to hearing from you.

With gratitude,

Pierre St. Clair

One Last Thing

If you liked this book, found it useful, enlightening, entertaining, or have something to say about it, could you post a short review on Amazon?

It doesn't have to be glowing, just your honest experience. When you share your comments it gives potential readers an idea of the book's value.

Here is the URL for readers in the USA:

http://amzn.to/2kOZMPD

Amazon will ask you to sign in and then take you directly to the review page where you can click on one of the stars to post your review.

If you do not live in the USA, you can post your review by clicking the review link on the same Amazon page where you purchased this book.

Thank you,

Pierre St Clair

ENDNOTES & BIBLIOGRAPHY

1   Kitty Ferguson, *Measuring the Universe*, New York: Walker and Company, 1999, p. 107

2   Fred Hoyle, *Astronomy and Cosmology*, San Francisco, W H Freeman and Co, 1975, p. 48

3   Paul Davies, *The Goldilocks Enigma*, Penquin Books, London, 2007, p. 9

4   James Jeans, *The Mysterious Universe*, Cambridge University Press, 1930, p.140

5   Mario Livio, *The Golden Ratio*, New York, Random House, 2002, p. 245

6   *Science at the Crossroads*, "Marx's Theory on the Historical Process," London, Frank Cass and Co., 1971, p. 189

7Ibid., p. 33

8   Karl Popper, *Conjectures and Refutations: The Growth of Scientific Knowledge*, p. 102

9   Morris Kline, *Mathematics: The Loss of Certainty*, Oxford University Press, 1980, p. 6

10   Tom van Flandern, *Dark Matter, Missing Planets and New Comets*, revised edition, Berkeley, CA: North Atlantic Books, 1993, p. xxi

11   "A Conversation with Brian Greene," Nova television series, PBS, October 2004

12   Bertrand Russell, *Mysticism and Logic*, Doubleday, 1957, pp. 70-71

13   Paul Davies, *The Goldilocks Enigma*, Penquin Books, London, 2007, p.129

14   CBC Television Interview - The Hour, 2007

15   www.youtube.com/watch?v=DfPeprQ7oGc

16   Alan Guth and Paul Steinhardt, "The Inflationary Universe," *Scientific American*, May 1984, p. 128

17   "Before the Big Bang There Was What?" *The New York Times*, May 22, 2001

18   Hawking and Mlodinow, *The Grand Design*, Bantam Books, New York, 2010, p. 207

19   Paul Davies, *The Goldilocks Enigma*, Penquin Books, London, 2007, p.86

20   Ronald W. Clark, *Einstein: The Life and Times*, Avon Books, New York, p. 301

21   Paul Davies, *The Goldilocks Enigma*, Penquin Books, London, 2007, p. 166

22   "Induction and Deduction in Physics," Berliner Tageblatt, December 25, 1919. Cited in *The Expanded Quotable Einstein*, p. 237

23   Robert Matthews, *New Scientist*, 30, 1, 1999, p. 24

24   Paul Davies, *The Goldilocks Enigma*, Penquin Books, London, 2007, p. 130

25   Philip and Phylis Morrison, "The Big Bang: Wit or Wisdom?" *Scientific American*, February 2001, p. 95

26   Mark Twain, *Life on the Mississippi*, 1883, p. 156

27   Karl Popper, Conjectures and Refutations: The Growth of Scientific Knowledge, p. 102

www.ingramcontent.com/pod-product-compliance
Lightning Source LLC
Chambersburg PA
CBHW021408170526
45164CB00002B/552